THE WOMAN'S DAY
Dictionary of
GLASS

Cover illustration: Goblet, Boston & Sandwich Glass Co.,
c. 1860-70, blown, stained, and engraved. Courtesy of The
Corning Museum of Glass, Corning, New York.

THE WOMAN'S DAY
Dictionary of
GLASS

General Editor: Dina von Zweck

Illustrations by Helen Disbrow
and Janet Hautau

A MAIN STREET PRESS BOOK

CITADEL PRESS
Secaucus, New Jersey

First edition

Text copyright © 1983 by The Main Street Press

Illustrations copyright © 1983 by CBS Publications, The
Consumer Publishing Division of CBS, Inc.

Published by Citadel Press
A division of Lyle Stuart, Inc.
120 Enterprise Ave., Secaucus, NJ 07094

In Canada: Musson Book Company
A division of General Publishing Co. Limited
Don Mills, Ontario

Produced by The Main Street Press

ISBN 0-8065-0841-8

Manufactured in the United States of America

Contents

Introduction

INTEREST in collecting glass made in Europe and North America from the 1700s to the 1930s grows with each passing year. More attention is paid to glass objects in today's annual price guides than to almost any other type of popular antique. Many millions of pieces of glass were produced in the past 300 years—from one-of-a-kind free-blown vases and pitchers to assembly-line-manufactured pressed goblets, tumblers, and plates. Some collectors look for particular patterns; others are interested only in the work of one designer or glasshouse; and still others concentrate on a special type such as lacy Sandwich glass, South Jersey glass, art glass, Depression glass, and brilliant cut glass. *The Woman's Day Dictionary of Glass* introduces the reader to the very wide variety of forms, designs, and makers that have marked the production of glass. The emphasis throughout is on American-made wares, since these are the most popular and easily collectible of antique glass objects.

The Woman's Day Dictionary of Glass presents examples of all major glass forms—from baskets and spoon holders to dishes modeled in the shape of various animals. Instructive background information on the development of thirty-three different forms—including such important categories as bottles, bowls, candlesticks, creamers, decanters, jugs and cruets, lamps, plates, sauce dishes, and sugar bowls—precedes each group of line drawings. Each of the more than 700 illustrated objects are further described and valued. The illustrations, many drawn from the manufacturers' own sales catalogs from the 1800s and early 1900s, capture the characteristic details of the pieces for easy identification. Particularly complex pieces are shown in larger scale than simpler objects.

Most glass made before the 1870s, whether blown or pressed, contains flint in its composition. After this time, it became common to substitute a lime-based formula for the flint, thereby lowering the manufacturing cost. Some glasshouses, particularly such well-established firms as the Boston & Sandwich Glass Co. and various of the Pittsburgh-area companies, continued to make both flint and non-flint wares. The weight of the latter pieces is lighter than those made of flint. While many collectors prefer the earlier flint pieces and will pay a high price for them, value is as much a reflection of the rarity of the pattern or design as it is of manufacture. The term "crystal" was generally used before the turn of the century to describe a brilliant clear color; by the early 1900s the term was reserved for a heavy glass, often cut in some manner, with a substantial percentage of lead in its composition.

Of all the types of antique glass produced, pressed glass far surpasses all others in quantity. Americans pioneered in the development of machinery to make pressing feasible on a large scale. They perfected the three basic steps of pressed glassmaking: the making of a wooden model and then a cast-iron or brass mold, the pressing of the molten glass by machinery, and finally its firing to a satisfactory brilliance. The value and rarity of pressed wares correspond quite directly to their age—from the limited early products of such pioneer firms as the Boston & Sandwich Glass Co., New England Glass Co., and Bakewell, Pears & Co., through the more extensive pattern wares of a multitude of Eastern and Midwestern firms in the mid- to late-1800s, and concluding with the nearly limitless production of Depression glass in the 1920s and '30s.

Free-blown and mold-blown glass objects are also included in these pages. Bottles and paperweights, for example, have commonly been blown rather than pressed. But in every category, there are exceptional objects which received special handling—cutting or shaping in an imaginative manner.

The value of a particular piece of glass will vary greatly in relation to condition, rarity, and demand. Colored pieces usually command prices 20% or so higher than clear pieces. Since value can vary so greatly, price ranges rather than set figures are provided for each illustrated object. These are coded A through H:

A = $2,500 and over
B = $1,000 to $2,500
C = $500 to $1,000
D = $250 to $500
E = $100 to $250
F = $50 to $100
G = $25 to $50
H = under $25

THE WOMAN'S DAY
Dictionary of
GLASS

1. Animal-Form Dishes

ANIMAL-form dishes, used for serving relishes and candy, and as holders for various household items, became extremely popular in the late 1800s. The fish, lion, turtle, duck, owl, lamb, swan, and turkey were among the most common forms to appear in the offerings of American pressed-glass manufacturers from the 1870s through the turn of the century. Porcelain serving dishes in various animal and vegetal forms were first popularized in the 18th century by such distinguished English and Continental porcelain firms as Chelsea and Meissen. Victorian glass manufacturers could press similar types at very low cost, therefore making them available to a wide segment of the population.

The finest and earliest Victorian pieces are those of opaque colored glass or milk glass and were produced by such firms as Atterbury & Co. (nos. 2 and 6) and Challinor, Taylor & Co. (no. 5), both of the Pittsburgh area. These figures usually feature ruby glass eyes, and, in addition to being made in white, were available in turquoise and an opalescent shade. Milk glass was first called opaque glass; now the term "milk glass" embraces such hues as blue and black, as well as milk white.

As animal-form dishes became more and more popular in the 1890s, clear or frosted pieces of a cheaper non-flint composition were widely produced, such as the turkey candy dish (no. 10) and various forms of fish relish dishes. These were mass-produced, and prices today are a great deal lower than for the earlier milk-glass varieties.

1. Relish dish, Atterbury & Co., Pittsburgh, 1870s, crystal, pressed. **(F)**
2. Covered dish, prob. Atterbury & Co., Pittsburgh, 1870s, ruby eyes, turquoise and opal milk glass body, pressed. **(F)**
3. Relish dish, Hobbs, Brockunier & Co., Wheeling, W. Va., 1870s, clear, pressed. **(F)**
4. Covered dish, maker unknown, 1880s, colored milk glass, pressed. **(E)**
5. Cream pitcher, Challinor, Taylor & Co., Ltd., Tarentum, Pa., 1880s, glass eyes, turquoise blue milk glass body, pressed. **(E)**
6. Covered dish, Atterbury & Co., Pittsburgh, 1880s, glass eyes, milk glass, pressed. **(E)**
7. Covered dish, maker unknown, 1880s, white milk glass lamb, turquoise blue fluted base, pressed. **(E)**
8. Covered dish, maker unknown, 1890s, white milk glass rabbit and basket base, pressed. **(G)**
9. Salt, maker unknown, early 1900s, black milk glass, pressed. **(G)**
10. Candy dish, prob. Cambridge Glass Co., Cambridge, Ohio, early 1900s, frosted glass, pressed. **(F)**

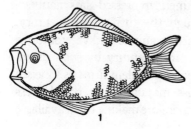

1

2

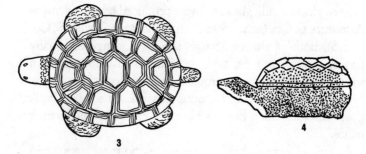

3

4

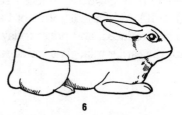

5

6

7

8

9

10

2. Baskets

BLOWN-molded or pressed-glass baskets are very charming objects. Because they are so appealing, they are difficult to find in the antiques marketplace and command relatively high prices. Baskets were commonly used in the 19th century for cakes or candies and were more often made of silver, Brittania metal, or porcelain or pottery.

Pressed-glass baskets simply do not have the light, frilly appearance of blown ware, whether Amberina or a solid color. But because they were not widely produced, their rarity has attracted the interest of collectors. Most of the varieties to be found today date from the 1880s through the early 1900s. Almost all, like nos. 1 and 3, have a pedestal base. Some, as in no. 1, have moveable metal handles rather than fixed, or applied glass handles that are blown or pressed. Baskets used for fruit or cake necessarily came either with a moveable handle or none at all.

---

1. Cake basket, Ripley & Co., Pittsburgh, 1880s, Mascotte pattern, pressed. **(F)**
2. Basket, prob. New England Glass Co., New Bedford, Mass., 1880s-90s, Amberina pattern, blown with molded handle. **(D)**
3. Cake basket, maker unknown, 1890s, Alexis pattern, pressed. **(G)**

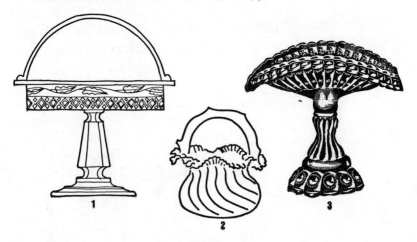

3. Bottles

Bottles come in every possible form, and collectors of them are legion. Since they were never considered very valuable objects, fine examples may be found buried in the backyard of an old house as frequently as in the display case of an antiques shop. Until the early 1800s, almost all bottles were free blown, and the mark of the pontil, the iron pipe used in blowing the form, is found on every variety of common blown bottle. As molds began to be used more and more in the 19th century for blown ware, free-blown production diminished greatly. Examples such as the wine bottle illustrated on the following page virtually disappeared, and bottle production became an assembly-line activity. The demand for containers to be used for medicinal purposes, alcoholic beverages, and sauces, grew and grew throughout the 1800s as commercial products replaced the homemade.

Among the most collectible of the early to mid-19th century mold-blown bottles are flasks (no. 3), figural whiskey bottles (nos. 8 and 9), and household bottles and castors (nos. 4, 5, and 7). The most commonly available bottles today are those for ink and patent medicine or pharmaceutical uses, such as those manufactured by the Whitall, Tatum Co. of Millville, N.J. from the 1850s through the early 1900s (nos. 13-27). Most of the examples shown are from the company's 1880 catalog of druggists', chemists', and perfumers' glassware and are without the final molded lettering (custom ordered by a druggist) or labels. The most valuable of the late blown-molded bottles are those which were cut (no. 28) or made for special commercial promotion (nos. 12, 30-31). The fanciest of all commercially-produced bottles are those for perfume and cologne, many of which were cut or engraved, and are described in section 23.

1. Wine bottle, prob. Wistarburg, N.J., stamped Wm Savery, 1752, dark green, free blown. **(A)**
2. Gin bottle, European or American, also known as a case bottle, late 1700s, mold blown. **(G)**
3. Violin flask, Pittsburgh, also known as scroll flask, 1840s-50s, blue-green, mold blown. **(E)**
4. Household bottle, New England, mid 1800s, clear, mold blown. **(E)**
5. Castor bottle, New England, mid 1800s, clear, with metal cap, Bull's Eye Variant pattern, mold blown. **(G)**
6. Barbershop bottle, New England, 1850s-60s, white overlay cut back to clear ruby, Leaf and Berry design, mold blown. **(F)**
7. Vinegar bottle, M'Kee & Bros., Pittsburgh, 1860s, clear, mold blown. **(G)**
8. Whiskey bottle, Whitney Glass Works, Glassboro, N.J., E. G. Booz Old Cabin design, 1860s-70s, light green, mold blown. **(A)**
9. Bitters bottle, patented by John B. Bartlett, New York, N.Y., 1869, Ear of Corn design, painted crystal, mold blown. **(E)**
10. Mustard bottle, King, Son & Co., Pittsburgh, 1870, metal cap, Plain design, mold blown. **(G)**

1

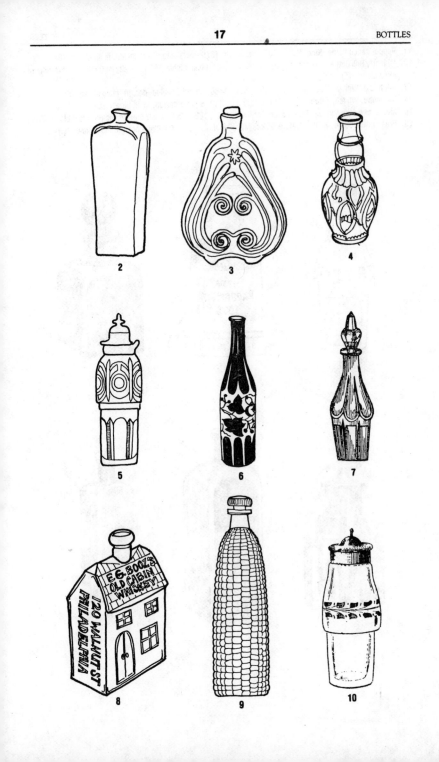

2

3

4

5

6

7

8

9

10

11. Pepper sauce bottle, King, Son & Co., Pittsburgh, 1870, metal cap, Plain design, mold blown. **(G)**
12. "Liberty Bell Bottle," patented by Samuel C. Vaughan, Philadelphia, 1874, novelty for Centennial Exhibition, mold blown. **(E)**
13. Ink bottle, Whitall, Tatum & Co., Millville, N.J., 1880, Round Shoulder design, mold blown. **(G)**
14. Ink bottle, Whitall, Tatum & Co., Millville, N.J., 1880, Square design, mold blown. **(G)**
15. Medicine bottle, Whitall, Tatum & Co., Millville, N.J., 1880, Tall French Square design, mold blown. **(H)**
16. Prescription bottle, Whitall, Tatum & Co., Millville, N.J., 1880, Tall Blake design, mold blown. **(H)**

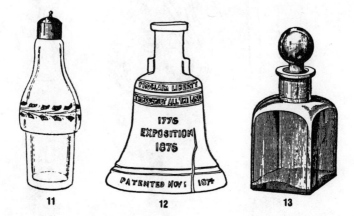

11   12   13

14   15   16

17. Tincture, Whitall, Tatum & Co., Millville, N.J., 1880, with Lubin stopper, mold blown. **(H)**
18. Saltmouth, Whitall, Tatum & Co., Millville, N.J., 1880, with Mushroom stopper, mold blown. **(H)**
19. Prescription bottle, Whitall, Tatum & Co., Millville, N.J., 1880, wide mouth design, mold blown. **(H)**
20. Prescription bottle, Whitall, Tatum & Co., Millville, N.J., 1880, New Philadelphia Oval design, mold blown. **(H)**
21. Cod liver bottle, Whitall, Tatum & Co., Millville, N.J., 1880, mold blown. **(H)**
22. Poison bottle, Whitall, Tatum & Co., Millville, N.J., 1880, mold blown. **(F)**

17

18

19

20

21

22

23. Bitters bottle, Whitall, Tatum & Co., Millville, N.J., 1880, Square Gothic design, mold blown. **(G)**
24. Pepper sauce bottle, Whitall, Tatum & Co., Millville, N.J., 1880, fluted design, mold blown. **(H)**
25. Nursing bottle, Whitall, Tatum & Co., Millville, N.J., 1880, Millville design, mold blown. **(F)**
26. Nursing bottle, Whitall, Tatum & Co., Millville, N.J., 1880, Acme design, mold blown. **(F)**
27. Nursing bottle, Whitall, Tatum & Co., Millville, N.J., 1880, Baltimore design, free blown. **(F)**
28. Worcestershire sauce bottle, T. B. Clark & Co., Honesdale, Pa., 1896, blown and cut. **(G)**

23

24

25

26

27

28

29. Mineral water bottle, New England, c. 1900, Moses with Staff design made for Poland Water, clear lavender, mold blown. **(F)**

30. Ink bottle, designed by Charles H. Henkels for Carter's Ink Co., Boston, 1914-16, Mr. Carter design, mold blown. **(F)**

31. Ink bottle, designed by Charles H. Henkels for Carter's Ink Co., Boston, 1914-16, Mrs. Carter design, mold blown. **(F)**

29

31

30

4. Bowls and Compotes

Bowls and compotes, originally termed comports, were standard items for European and American glass manufacturers in the 1700s and 1800s. Whether for preparing or serving cold food, a glass bowl has an appeal which outweighs that of metal or clay. Rarely, however, is glass used for hot dishes unless it has been tempered in some manner, as is commonly the practice today. In size, bowls range from tiny berry and finger bowls of 5" diameter to large punch bowls measuring 14" to 15". The earliest products from European and American glasshouses were free blown, and among the most attractive and simple are the folded-rim vessels of the late 1700s and early 19th century (nos. 1 and 2). Almost as prized and expensive are the cut varieties produced by such late-Victorian American firms as T. B. Clark and Dorflinger in Honesdale, Pa. (nos. 38-45). Pressed-glass imitations of these pieces were widely manufactured.

The catalogs of 19th-century American firms often refer to a compote, a bowl joined to a base. It was easy for the pressed-glass manufacturer to interchange bowl and base. The same bowl design might be used with a high-footed or low-footed section. High-footed compotes were often used for serving fruit; the lower might serve as sauce or jelly dishes. By the late 1800s, the compote form lost much of its popularity, and its place was taken by such smaller-scale specialized dishes as the individual berry bowl and the salad bowl. Large serving bowls continued to be made, of course, but like most of those manufactured in the 1820s and '30s, they are without a footed base.

1. Bowl, maker unknown, Zanesville, Ohio area, early 19th century, blown with folded rim. **(C)**
2. Bowl, maker unknown, prob. New York City area, c. 1830, aquamarine, blown. **(D)**
3. Bowl, Bohemia, c. 1830, wheel engraved and facet-cut. **(C)**

4. Bowl, Redwood Glass Works, Redwood, N.J., c. 1833-50, lily pad decoration, blown. **(C)**
5. Bowl, Waterford Glass Works, Waterford, Ireland, 1840s, Lozenges design, cut. **(E)**
6. Bowl, New England Glass Co., East Cambridge, Mass., c. 1840, mold blown. **(E)**
7. Cracker bowl, M'Kee & Bros., Pittsburgh, 1864, pressed. **(G)**
8. Covered bowl, M'Kee & Bros., Pittsburgh, 1868, Eureka pattern, pressed. **(F)**
9. Bowl, M'Kee & Bros., Pittsburgh, 1868, New Pressed Leaf pattern, pressed. **(F)**

10. Bowl, M'Kee & Bros., Pittsburgh, 1868, Concave pattern, pressed. **(G)**
11. Compote, King, Son & Co., Pittsburgh, 1870, plain ware, pressed. **(G)**
12. Compote, King, Son & Co., Pittsburgh, 1870, scalloped edge, pressed. **(G)**
13. Bowl, King, Son & Co., Pittsburgh, 1870, Jewel pattern, pressed. **(F)**
14. Bowl, M'Kee & Bros., Pittsburgh, 1871, Sprig pattern, pressed. **(G)**
15. Covered bowl, M'Kee & Bros., Pittsburgh, 1871, plain design, pressed. **(H)**

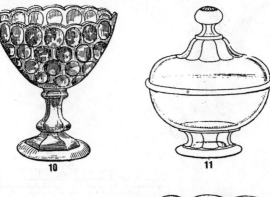

10

11

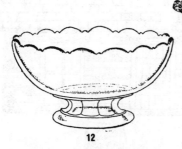

12

13

14

15

16. Bowl, M'Kee & Bros., Pittsburgh, 1871, Leaf pattern, pressed. **(G)**
17. Covered compote, Central Glass Co., Wheeling, W. Va., c. 1875, Log Cabin pattern, pressed. **(E)**
18. Compote, J. P. Higbee Glass Co., Bridgeville, Pa., early 1900s, Paneled Thistle pattern, pressed. **(G)**

16

17

18

19. Dish, Campbell, Jones & Co., Pittsburgh, 1880s, Rose Sprig pattern, pressed. **(G)**
20. Coach bowl, M'Kee & Bros., Pittsburgh, patented 1886 by Julius Praeger, simulated wheel supports, pressed. **(G)**
21. Covered bowl, Dalzell, Gilmore & Leighton Glass Co., Findlay, Ohio, c. 1889, lustre floral Onyx design, mold blown. **(F)**
22. Berry bowl, maker unknown, 1890s, Russian pattern, cut. **(F)**
23. Berry bowl, Richards & Hartley Flint Glass Co., Pittsburgh, 1890s, Heart with Thumbprint pattern, pressed. **(G)**
24. Covered berry bowl, Fostoria Glass Co., Moundsville, W. Va., 1890s, St. Bernard pattern, pressed and engraved. **(F)**

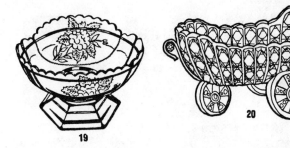

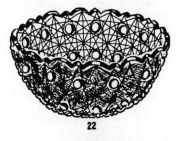

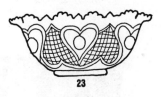

25. Rose bowl, prob. M'Kee & Bros., Pittsburgh, c. 1894, Queen pattern, pressed. **(G)**
26. Finger bowl, prob. M'Kee & Bros., Pittsburgh, c. 1893, Majestic pattern (also called Magic or Puritan), pressed. **(H)**
27. Finger bowl, Fostoria Glass Co., Moundsville, W. Va., 1895, Czarina pattern, pressed. **(H)**
28. Fruit bowl, maker unknown, 1890s, pressed. **(H)**
29. Salad bowl, maker unknown, 1890s, strawberry diamond and fan design, cut. **(F)**
30. Berry or fruit bowl, maker unknown, 1890s, strawberry diamond and fan design, cut. **(F)**
31. Berry dish, maker unknown, 1890s, Alexis pattern, pressed. **(H)**
32. Fruit or cake bowl, maker unknown, 1890s, ruby and crystal glass, pressed. **(G)**
33. Berry dish, maker unknown, 1890s, leaf design, engraved. **(H)**

25

26

27

28

29

30

31

32

33

34. Berry or fruit dish, maker unknown, 1890s, strawberry diamond and fan design, pressed. **(H)**

35. Berry or fruit dish, maker unknown, 1890s, Princeton pattern, pressed. **(H)**

36. Berry bowl, maker unknown, 1890s, Pittsburgh area, Jeannette pattern, pressed. **(H)**

37. Berry dish, Fostoria Glass Co., Moundsvile, W. Va., 1895, Czarina pattern, pressed. **(H)**

38. Finger bowl, T. B. Clark & Co., Honesdale, Pa., 1896, Winola pattern, cut. **(F)**

39. Ice Tub, T. B. Clark & Co., Honesdale, Pa., 1896, Jewel pattern, cut. **(E)**

40. Punch bowl, T. B. Clark & Co., Honesdale, Pa., 1896, Desdemona pattern, cut. **(D)**

41. Bowl, T. B. Clark & Co., Honesdale, Pa., 1896, Arbutus pattern, cut. **(E)**

34

35

36

37

38

39

40

41

42. Bowl, T. B. Clark & Co., Honesdale, Pa., 1896, Adonis pattern, cut. **(E)**
43. Bowl, T. B. Clark & Co., Honesdale, Pa., 1896, Manhattan pattern, cut. **(E)**
44. Bowl, T. B. Clark & Co., Honesdale, Pa., 1896, Palmetto pattern, cut. **(E)**
45. Rose globe, T. B. Clark & Co., Honesdale, Pa., 1896, Manhattan pattern, cut. **(E)**
46. Fruit bowl, U. S. Glass Co., Pittsburgh, 1904, crimped edges, pressed. **(G)**
47. Fruit bowl, U. S. Glass Co., Pittsburgh, 1904, New Jersey pattern, pressed. **(G)**

48. Compote, maker unknown, early 1900s, leaf design, pressed with engraving. **(H)**
49. Bowl, U. S. Glass Co., Pittsburgh, 1904, Pennsylvania pattern, pressed. **(H)**
50. Punch bowl, U. S. Glass Co., Pittsburgh, 1904, Pennsylvania pattern, pressed. **(F)**
51. Compote, Indiana Glass Co., Dunkirk, Ind., late 1920s, Old English pattern, pressed. **(H)**
52. Bowl, L. E. Smith Co., 1920s-30s, Double Shield pattern, black, pressed. **(H)**
53. Bowl, Hocking Glass Co., Lancaster, Ohio, late 1930s, Fortune pattern, pink, pressed. **(G)**

5. Butter Dishes

Dishes used for serving butter at the table are among the most highly decorative of glass objects. They bear little or no resemblance to the strictly utilitarian rectangles which are used in the refrigerator today and sometimes make their way to the table. In the days before butter was bought in neatly packaged sticks, butter was cut into squares or scooped out from a tub much in the way ice cream is now. The traditional butter dish, then, is a circular piece. It consists of two pieces, a saucer and cover, and sometimes includes a pierced strainer on which the butter rests.

Pressed-glass butter dishes were commonly used at the table during the 1800s and were often a standard part of a pattern set. Typical of these dishes are the Ruby Rosette (no. 1), Chrysanthemum Leaf (no. 3), Roman Rosette (no. 5), Flower and Panel (no. 6), St. Bernard (no. 7), and Pomona (no. 12) pattern pieces. Even more imaginative are such decorative dishes as the Cabbage Leaf (no. 2) and Leaf (no. 8) examples.

Almost all the antique glass butter dishes that one is likely to discover today date from the second half of the 19th century. Earlier plain and lacy Sandwich and Pittsburgh varieties do exist, but often either the saucer or the cover has not survived. Prior to the 1800s, pottery or porcelain was more commonly used for such specialty dishes.

1. Butter dish, Bryce, M'Kee & Co., Pittsburgh, late 19th century, Ruby Rosette pattern, pressed. **(G)**
2. Butter dish, maker unknown, late 19th century, Cabbage Leaf pattern, pressed. **(F)**
3. Butter dish, maker unknown, late 19th century, Chrysanthemum Leaf pattern, pressed. **(F)**
4. Butter dish, King, Son & Co., Pittsburgh, 1870s, flange design, pressed. **(G)**

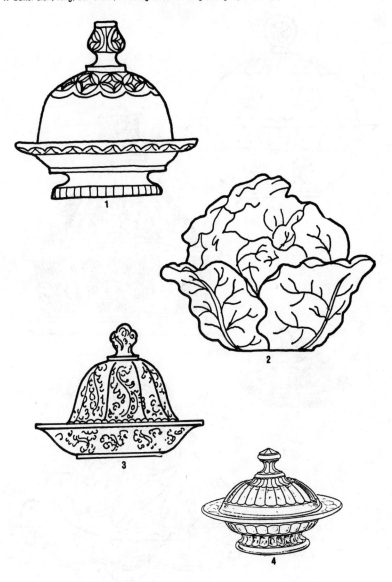

5. Butter dish, Bryce, M'Kee & Co., Pittsburgh, c. 1875, Roman Rosette pattern, pressed. **(G)**

6. Butter dish, Challinor, Taylor & Co., Tarentum, Pa., c. 1885, Flower and Panel pattern, pressed. **(G)**

7. Butter dish, Fostoria Glass Co., Moundsville, W. Va., 1890s, St. Bernard pattern, pressed. **(F)**

8. Butter dish, prob. J. H. Hobbs Glass Co., Wheeling, W. Va., 1890s, Leaf pattern, pressed. **(E)**

9. Butter dish, prob. U. S. Glass Co., Pittsburgh, 1890s, Hero ruby engraved pattern, pressed. **(G)**
10. Butter dish, maker unknown, 1890s, Alexis pattern, pressed. **(G)**
11. Butter dish, Riverside Glass Co., Wellsburg, W. Va., 1890s, Ray pattern, pressed. **(G)**
12. Butter dish, maker unknown, late 19th century, Pomona pattern, pressed. **(G)**
13. Butter tub and plate, T. B. Clark & Co., Honesdale, Pa., 1890s, Manhattan pattern, cut. **(E)**
14. Cheese cover and plate, T. B. Clark & Co., Honesdale, Pa., 1890s, Manhattan pattern, cut. **(D)**
15. Butter dish, Hazel Atlas Glass Co., 1930s, Wedding Band pattern, pressed and molded, with metal cover. **(G)**

9

10

11

12

13

14

15

6. Candlesticks

Since most candlesticks are made of metal, pottery, or wood, those of glass are especially valuable and collectible. They were rarely offered as part of pattern glass sets, but were purchased individually. The most sought-after are the colorful early Sandwich and Pittsburgh pressed sticks dating from the 1830s through the '70s. The most common forms are the columnar shape, such as those illustrated in nos. 5 and 6; the dolphin and petticoat dolphin forms (nos. 3 and 4); and the crucifix form (no. 2). Many glasshouses in New England and western Pennsylvania produced these types. The colors include electric blue, cobalt blue, deep amethyst, and emerald green, with canary yellow being the most common. During the 1840s and '50s such opaque shades as grease blue, clambroth, and a creamy white were popular.

With the exception of the 18th-century South Jersey candlestick illustrated in no. 1, all those shown here are completely pressed. The collector, however, may encounter sticks that were assembled from pressed and blown or blown-molded elements, and such combined forms usually date from the early 1800s.

Depression-glass candlesticks from the 1920s and '30s are often without a shaft. Typical of this chamber stick form is the Diamond Quilted pattern type made by the Imperial Glass Co. Chamber sticks of an earlier period are usually miniature versions of larger pieces, the socket, shaft, and base being merely scaled down.

1. Candlestick, prob. Wistarburg, N.J., c. 1740-80, light green, free blown. **(C)**
2. Candlestick, Boston & Sandwich Glass Co., 1860s, opaque white, cruciform, hexagonal base, pressed. **(D)**

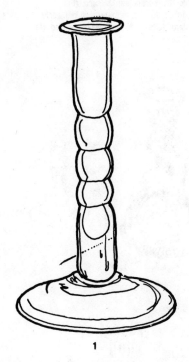

1

2

3. Candlestick, M'Kee & Bros., Pittsburgh, 1860s-70s, dolphin form, pressed. **(E)**
4. Candlestick, maker unknown, Pittsburgh, 1860s-70s, petticoat dolphin form, pressed. **(E)**
5. Candlestick, M'Kee & Bros., Pittsburgh, 1860s-70s, French form, pressed. **(F)**
6. Candlestick, M'Kee & Bros., Pittsburgh, 1860s-70s, Boston form, pressed. **(F)**
7. Chamber stick, King, Son & Co., Pittsburgh, 1870s, Jewel pattern, pressed with applied handle. **(G)**
8. Candlesticks, maker unknown, prob. Midwestern, 1880s, white milk glass, pressed. **(F)**

3

4

5

6

7

8

9. Candlestick, maker unknown, prob. Midwestern, 1890s, Pillar pattern, pressed. **(G)**
10. Candlestick, U. S. Glass Co., Pittsburgh, 1904, Pennsylvania pattern, pressed. **(G)**
11. Candlestick, Indiana Glass Co., Dunkirk, Ind., late 1920s, Old English pattern, pressed. **(H)**
12. Candlesticks, Imperial Glass Co., Bellaire, Ohio, late 1920s-early 1930s, Diamond Quilted or Flat Diamond pattern, pressed. **(G)**

7. Celery Vases

CELERY vases or holders, commonly used in the 1800s, are often mistaken for goblets or small flower vases. The practice of serving celery on a tray with other hors d'oeuvres is primarily a 20th-century phenomenon. The height of the average celery, nine to ten inches, closely approximates that of a flower vase, but the stem on the celery is shorter and may resemble that found on a goblet. The bowl of a celery, however, is usually somewhat fuller and less tapered than that of a goblet.

Celery vases appear to have been made first in England and Ireland during the late 18th century and were practically without a stem or shaft. American glassmakers introduced the goblet form in the mid-1800s, and a celery was often offered as part of a pattern glass set. Representative of this pressed-glass type are the Honeycomb (no. 2), Crystal (no. 3), Huber (no. 5), Flat Diamond (no. 6), and Thousand Eye (no. 7) pattern pieces. Later 19th-century and early 20th-century celery vases are usually found without a stem (nos. 8-15).

Colored celery vases are a great rarity in the wares of the mid-1800s; examples in amber, blue, apple green, and vaseline are more likely to be found in later pattern glass.

1. Celery vase, Boston & Sandwich Glass Co., 1820s-30s, Diamond Quilting pattern with ribbing, blown molded. **(C)**
2. Celery vase, maker unknown, Pittsburgh area, 1860s, Honeycomb pattern, pressed. **(G)**
3. Celery vase, M'Kee & Bros., Pittsburgh, 1860s, Crystal pattern, pressed. **(G)**
4. Celery vase, King, Son & Co., Pittsburgh, 1870s, Gothic pattern, pressed. **(F)**

1

2

3

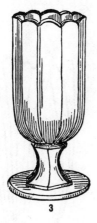

4

5. Celery vase, M'Kee & Bros., 1870s, Huber pattern, pressed. **(G)**
6. Celery vase, Richard & Hartley Flint Glass Co., Pittsburgh, Flat Diamond pattern, pressed. **(G)**
7. Celery vase, prob. Adams & Co., Pittsburgh, 1870s, Thousand Eye pattern, pressed. **(F)**
8. Celery vase, Adams & Co., Pittsburgh, 1870s, Plume pattern, pressed. **(G)**
9. Celery vase, maker unknown, prob. Midwest, 1870s, Millard pattern, pressed. **(G)**

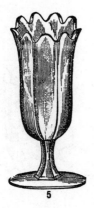

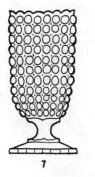

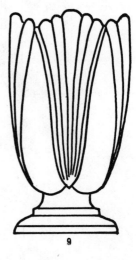

10. Celery vase, U. S. Glass Co., Pittsburgh, 1890s, Kentucky pattern, pressed. **(H)**
11. Celery vase, prob. Hobbs, Brockunier & Co., Wheeling, W. Va., 1890s, Optic pattern, pressed. **(H)**
12. Celery vase, maker unknown, 1890s, Alexis pattern, pressed. **(H)**
13. Celery vase, U. S. Glass Co., Pittsburgh, 1904, Manhattan pattern, pressed. **(H)**
14. Celery vase, U. S. Glass Co., Pittsburgh, 1904, Pennsylvania pattern, pressed. **(H)**
15. Celery holder, U. S. Glass Co., Pittsburgh, 1904, Virginia pattern, pressed. **(H)**

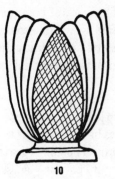

10

11

12

13

14

15

8. Creamers

CREAM pitchers are widely collected today for both their utility and beauty. Their basic form has changed little from the time when such beverages as coffee and tea became common in the American household over 250 years ago. While cream may not be used as often today for desserts and drinks, the matching creamer and sugar set is still considered essential for a properly laid table.

Creamers are always handled and are often bulbous in form with a low center of gravity, making them less subject to upset. Some of the 18th-century blown varieties have applied feet; a circular flat base is the rule for pressed forms. A creamer was a standard part of the mid- to late-19th century pattern-glass set and was available in a wide variety of colors. The decoration became more and more elaborate throughout the Victorian era. The most expensive and highly decorative of late 19th-century creamers are those of cut glass.

Pressed-glass creamers were sometimes made in two sizes in the same pattern—one for use at the dinner table, and a second for a breakfast set. Handles on the earlier pressed pieces (1830s-70s) are likely to be mold-blown and applied; later creamers may have handles which were pressed along with the body.

1. Creamer, prob. Boston & Sandwich Glass Co., mid-19th century, Star and Punty pattern, clear, pressed with applied molded handle. **(E)**
2. Creamer, Bakewell, Pears & Co., Pittsburgh, mid-19th century, Belted Icicle pattern, clear, pressed with applied molded handle. **(F)**
3. Creamer, prob. Boston & Sandwich Glass Co., 1840s-50s, Peacock Feather pattern, pressed with applied molded handle. **(E)**
4. Creamer, Franklin Flint Glass Co., Philadelphia, 1860s, Honeycomb pattern, pressed with applied molded handle. **(G)**

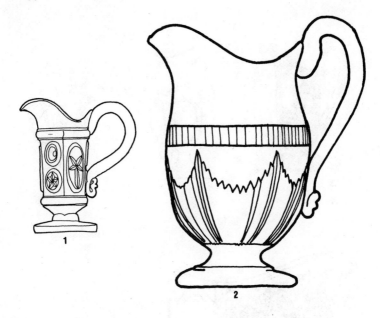

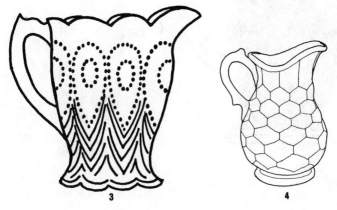

5. Creamer, M'Kee & Bros., Pittsburgh, 1868, Sprig pattern, pressed with applied molded handle. **(H)**
6. Creamer, maker unknown, c. 1870s, Loop and Moose Eye pattern, rare flint glass, pressed with applied molded handle. **(G)**
7. Creamer, prob. Atterbury & Co., Pittsburgh, 1870s, Cameo pattern, pressed with applied molded handle. **(F)**
8. Creamer, Portland Glass Co., Portland, Me., c. 1870, Tree of Life pattern, dark blue, pressed with applied molded handle. **(G)**
9. Creamer, Boston & Sandwich Glass Co., c. 1870, Ribbed Ivy pattern, pressed with applied molded handle. **(E)**

5

6

7

8

9

10. Creamer, Hartley Glass Co., Tarentum, Pa., 1880s, Block and Fan pattern, clear, pressed. **(H)**

11. Creamer, prob. Doyle & Co., Pittsburgh, 1880s, Red Block pattern, pressed with applied molded handle. **(G)**

12. Creamer, prob. Dithridge & Co., Pittsburgh, late 19th century, Cosmos pattern, milk white, pressed with applied molded handle. **(E)**

13. Creamer, maker unknown, late 19th century, Cardinal pattern, pressed with applied molded handle. **(G)**

10

11

12

13

14. Creamer, prob. Pittsburgh area, 1890s, Excelsior pattern, ruby, pressed with applied molded handle. **(F)**
15. Creamer, maker unknown, 1890s, strawberry, diamond, and fan design, cut. **(F)**
16. Creamer, Fostoria Glass Co., Moundsville, W. Va., 1895, Czarina pattern, pressed with applied molded handle. **(G)**
17. Creamer, A. H. Heisey Glass Co., Newark, Ohio, 1897, Hobstar pattern, pressed with applied molded handle. **(H)**
18. Creamer, U. S. Glass Co., Pittsburgh, 1904, Pennsylvania pattern, pressed with applied molded handle. **(H)**
19. Creamer, U. S. Glass Co., Pittsburgh, 1904, Wyoming pattern, pressed with applied molded handle. **(G)**
20. Creamer, prob. Pittsburgh area, 1927-32, Round Robin pattern, amber, pressed with applied molded handle. **(H)**
21. Creamer, Federal Glass Co., 1937-41, Diana pattern, amber, pressed with applied molded handle. **(H)**

14

15

16

17

18

19

20

21

9. Cup Plates

Cup plates are delightful small saucers which were once widely used in the drinking of tea. Odd as it may seem, until the mid-1800s, tea was often poured into a saucer to cool and then sipped from it rather than from a cup. A cup plate or saucer was a much more important object to decorate than a cup.

The New England glasshouses such as the Boston & Sandwich Glass Co. and the New England Glass Co. were among the earliest manufacturers of cup plates, beginning in the 1820s. At the same time, Pittsburgh-area glassmakers were improving pressing techniques, and cup plates were among their first products. There are over a thousand designs which may be attributed to several dozen manufacturers from the period 1820-50. In most reference books, the patterns are broken down into historical or representational and geometric categories. Ruth Webb Lee and James H. Rose wrote the bible on the subject, *American Glass Cup Plates*, in 1948, and many collectors find it useful to refer to the Lee and Rose pattern numbers.

The usual cup plate measures from two to three inches in diameter and has a scalloped border. Geometric designs are more commonly found than those of historical interest, such as the William Henry Harrison bust (no. 5) or such scenes as that shown in illustration no. 2. Clear examples are found more easily than colored and, as in other categories of glass, are lower in price. The lowly cup plate, originally sold for as little as 5¢, can now bring as much as $150.

1. Cup plate, maker unknown, prob. New England or Pennsylvania, 1820s, central star design, pressed. **(F)**
2. Cup plate (also called Fulton Plate), maker unknown, prob. Pittsburgh area, c. 1835-38, design of paddle wheel steamer, clear, pressed. **(E)**
3. Cup plate, maker unknown, c. 1830-45, center boss motif, scalloped rim, pressed. **(F)**
4. Cup plate, prob. Boston & Sandwich Glass Co., 1835-50, central geometric motif with border of hearts, blue, pressed. **(E)**

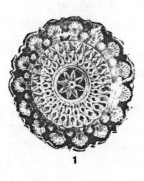

1

2

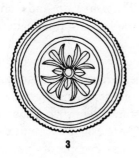

3

4

5. Commemorative cup plate, Boston & Sandwich Glass Co., 1841, William Henry Harrison portrait bust, pressed. **(G)**
6. Cup plate, prob. Boston & Sandwich Glass Co., mid 1800s, lacy design, pressed. **(F)**
7. Cup plate, prob. Boston & Sandwich Glass Co., mid 1800s, Lacy Hearts pattern, pressed. **(E)**
8. Cup plate, prob. Boston & Sandwich Glass Co., mid 1800s, star or rosette central motif, pressed. **(F)**

5

6

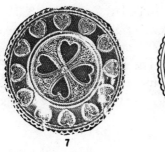

7

8

10. Cups and Dessert Cups

G LASS cups have been traditionally used for serving cold drinks such as punch or lemonade, or desserts such as custard or sherbet. Those for drinking are not footed or stemmed, of course; dessert cups, however, may resemble drinking cups or assume the pedestal form. Both types of cups might be supplied with matching saucers.

Sherbet was one of the favorite desserts of the late-Victorian household, and cups for this purpose were made by almost all of the major American manufacturers of pressed glass. Representative of this type are the Hobnail (no. 2), Chrysanthemum Leaf (no. 3), Adams (no. 4), Daisy and Cube (no. 5), and Triple Triangle (no. 6) pattern pieces. Among the most attractive of the punch or lemonade cups are two of special manufacture: cut glass varieties, such as that produced in the 1890s by T. B. Clark & Co. (no. 13); and art glass, both mold blown and pressed, Amberina cups such as the example (no. 7) probably made by the New England Glass Co. in the 1880s.

Footed dessert cups for ice cream, custard, and fruit became popular in the early 1900s and were made by most of the Depression-glass firms. The Imperial Glass Co.'s Diamond Quilted cup and saucer (no. 16) and Federal Glass Co.'s Rear Optic pattern cup (no.17) are typical of these colored wares.

1. Dessert dish, Richardson, Stourbridge, England, 1850s, pressed. **(G)**
2. Sherbet cup, prob. Adams & Co., Pittsburgh, 1870s-80s, Hobnail pattern, pressed. **(G)**
3. Sherbet cup, Boston & Sandwich Glass Co., 1880s, Chrysanthemum Leaf pattern, pressed. **(G)**
4. Sherbet cup, Adams & Co., Pittsburgh, 1880s, Adams pattern, pressed. **(G)**
5. Sherbet cup, maker unknown, 1880s, Daisy and Cube pattern, pressed. **(H)**
6. Sherbet cup, Doyle & Co., Pittsburgh, 1880s, Triple Triangle pattern, pressed. **(H)**

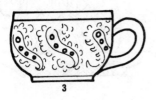

1

2

3

4

5

6

7. Punch cup, prob. New England Glass Co., E. Cambridge, Mass., 1880s, Inverted Thumbprint or Polka Dot pattern, clear amberina red shading to yellow, pressed. **(E)**

8. Dessert cup and saucer, Challinor, Taylor & Co., Ltd., Tarentum, Pa., Scroll with Star pattern, pressed. **(G)**

9. Dessert cup and saucer, U. S. Glass Co., Pittsburgh, 1890s, Panel and Flute pattern, pressed. **(F)**

10. Dessert cup and saucer, Bryce, M'Kee & Co., Pittsburgh, 1890s, Berkeley pattern, pressed. **(G)**

11. Sherbet cup, maker unknown, Pittsburgh area, 1890s, Diamond pattern, pressed. **(G)**

12. Custard or dessert glass, maker unknown, 1890s, pressed. **(H)**

7

8

9

10

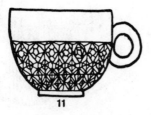

11

12

13. Lemonade cup, T. B. Clark & Co., Honesdale, Pa., Winola pattern, cut. **(G)**
14. Lemonade or custard glass, U. S. Glass Co., Pittsburgh, 1904, Pennsylvania pattern, pressed. **(H)**
15. Custard glass, U. S. Glass Co., Pittsburgh, 1904, Colorado pattern, pressed. **(H)**
16. Dessert glass and saucer, Imperial Glass Co., Bellaire, Ohio, late 1920s-early 1930s, Diamond Quilted pattern, pressed. **(H)**
17. Sherbet cup, Federal Glass Co., 1929-30, Rear Optic pattern, pressed. **(H)**
18. Dessert cup and saucer, Federal Glass Co., 1929-33, Optic, Raindrop pattern, pressed. **(H)**
19. Dessert cup and saucer, Jeannette Glass Co., Jeannette, Pa., 1929-33, Cube, Cubist pattern, pressed. **(H)**
20. Sherbet cup, Hazel Atlas Glass Co., 1936-40, Newport, Hairpin pattern, pressed. **(H)**

13   14   15

16

17

18

19

20

11. Decanters and Carafes

Decanters and carafes are fancy bottles for holding wine or other spirits; carafes are also used for drinking water. The decanter with a stopper is the larger of the two vessels and may hold a good supply of alcohol. The unstoppered carafe is often filled for each meal and placed at the table with two goblets; it was also used with a glass in the bedroom on a night stand. All the examples shown in the following pages are free blown or mold blown.

Decanters are among the most elegant of glass objects, and the decoration reflects their sophisticated use. Stoppers are usually of cut glass, and, if the body is not cut or engraved, it is commonly patterned in some other handsome manner. One of the fanciful forms is the calabash (no. 2), blown in a 24-rib mold and often found in a light-blue or green shade. The Horn of Plenty and Chain pattern decanters (nos. 4 and 5, respectively), are blown three mold, and each is shown with its original stopper.

Cut-glass decanters and carafes command high prices. In the late 1800s, English, Irish, and American-made cut glass were considered a particularly splendid wedding or anniversary gift. The usual decanter was quart size (no. 18), but a pint size in the same pattern was commonly offered as well. Carafes were also supplied in the two sizes. Many carafes, cut and uncut, were made for the hotel market, and they remain today more plentiful and lower in price than decanters.

1. Decanter, maker unknown, prob. Pennsylvania, 1770-90, later cut stopper, free blown. **(D)**
2. Calabash, maker unknown, Midwestern, prob. Zanesville, Ohio, c. 1815, mold blown. **(D)**
3. Quart decanter, Boston & Sandwich Glass Co., 1830s-40s, Horn of Plenty pattern, mold blown. **(E)**
4. Quart decanter, Boston & Sandwich Glass Co., 1830s-40s, Chain pattern, mold blown. **(E)**

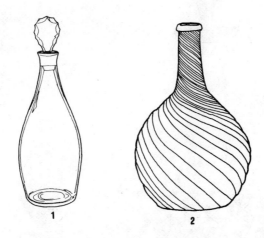

1

2

3

4

5. Decanter, maker unknown, English, 1851, cut. **(D)**
6. Decanter, maker unknown, English, 1851, cut. **(D)**

5

6

7. Claret decanter, maker unknown, English, 1851, blown. **(E)**
8. Wine decanter, maker unknown, English, 1851, mold blown. **(E)**
9. Decanter, maker unknown, English, 1851, cut. **(D)**
10. Decanter, Boston & Sandwich Glass Co., mid 1800s, Sandwich Star pattern, pressed and blown. **(D)**
11. Quart decanter, M'Kee & Bros., Pittsburgh, 1860s, Ribbed Leaf or Bellflower pattern, mold blown. **(E)**
12. Bar Bottle, King, Son & Co., Pittsburgh, 1870s, No. 138 pattern, mold blown. **(F)**

13. Decanter, maker unknown, 1870s-80s, Wedding Ring pattern, mold blown. **(F)**

14. Decanter, prob. George Duncan & Sons, Pittsburgh, c. 1880s-90s, Roman pattern, mold blown. **(F)**

15. Quart carafe, T.B. Clark & Co., Honesdale, Pa., 1890s, Winola pattern, cut. **(E)**

16. Quart carafe, T.B. Clark & Co., Honesdale, Pa., 1890s, Manhattan pattern, cut. **(E)**

13

14

15

16

17. Priscilla carafe, T.B. Clark & Co., Honesdale, Pa., 1890s, Manhattan pattern, cut. **(E)**
18. Quart no handle decanter, T.B. Clark & Co., Honesdale, Pa., 1890s, Winola pattern, cut. **(E)**
19. Wine decanter, prob. A.H. Heisey Glass Co., Newark, Ohio, c. 1897, Hobstar pattern with other motifs, mold blown. **(F)**
20. Carafe, prob. A.H. Heisey Glass Co., Newark, Ohio, c. 1897, Hobstar pattern with other motifs, mold blown. **(F)**
21. Water carafe, U.S. Glass Co., Pittsburgh, early 1900s, Virginia pattern, pressed. **(H)**
22. Decanter, Hazel Atlas Glass Co., 1930-35, New Century pattern, mold blown. **(H)**

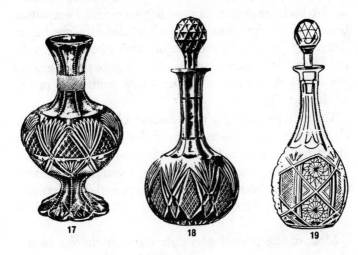

12. Dishes

"**D**ISHES" is a word used today to describe a wide variety of objects—plates, platters, bowls, cups and saucers, tureens—used in both the preparation and serving of food. Nineteenth-century glassmakers turned out such articles, of course, but, in describing pieces in their catalogues, they often reserved the label "dish" for objects which were likely to be passed around the table—serving dishes for vegetables, fruit, nuts, candy, relish, and jelly.

The most interesting of these dishes took unusual forms, including such examples as a lacy shell-shaped dish from the 1830s (no. 1), a shell pickle dish made by M'Kee & Bros. in the 1860s (no. 2), a swan finial oval covered dish dating from the 1880s (no. 14), and a Dewey pattern covered dish honoring Admiral George Dewey after his victory at Manila Bay in 1898 (no. 16).

Most of the pressed-glass dishes are far simpler serving pieces, covered or uncovered, in 7″, 8″, and 9″ sizes. A few of them are footed or stand on a pedestal. The term "nappy" was used in the late 1800s to describe a flat-bottomed dish with slightly curving sides. It might be used for sauces, relish, or berries, and have an attached handle, as in no. 12. Bonbon dishes were often made of cut blown glass and by the 1890s had become an essential decorative furnishing in the fashionable parlor. Illustrated are several exquisite examples by T. B. Clark & Co. of this type of dish intended for candy and small cakes (nos. 17-20).

1. Shell-shaped dish, maker unknown, New England, 1830-35, lacy, Peacock-Eye design, pressed. **(G)**
2. Shell-shaped pickle dish, M'Kee & Bros., Pittsburgh, 1860s, pressed. **(G)**
3. Oval dish, M'Kee & Bros., Pittsburgh, 1860s, Star pattern, pressed. **(G)**

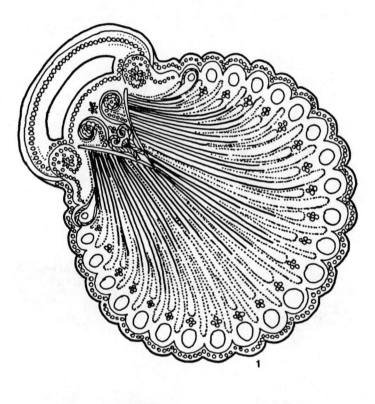

1

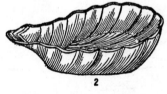

2

3

4. Nappy, M'Kee & Bros., Pittsburgh, 1860s, Ray pattern, pressed. **(G)**
5. Dish, M'Kee & Bros., Pittsburgh, 1860s, Ray pattern, pressed. **(G)**
6. Dish, M'Kee & Bros., Pittsburgh, late 1860s, Eureka pattern, pressed. **(G)**
7. Dish, M'Kee & Bros., Pittsburgh, late 1860s, Sprig pattern, pressed. **(H)**
8. Footed, covered dish, M'Kee & Bros., late 1860s, Crystal pattern, pressed. **(G)**
9. Footed dish, King, Son & Co., Pittsburgh, 1870s, Centennial pattern, pressed. **(G)**

10. Footed, covered oval dish, King, Son & Co., Pittsburgh, 1870s, Mitchell C. Ware, pressed. **(H)**
11. Footed, covered nappy, King, Son & Co., Pittsburgh, 1870s, Centennial pattern, pressed. **(G)**
12. Nappy, King, Son & Co., Pittsburgh, 1870s, Jewel pattern, pressed. **(H)**
13. Fruit dish, maker unknown, 1880s-90s, milk glass, pressed. **(H)**
14. Oval covered dish, prob. Canton Glass Co., Canton, Ohio, c. 1882, swan finial, pressed. **(F)**
15. Footed jelly dish, Fostoria Glass Co., Fostoria, Ohio, c. 1890, St. Bernard or No. 450 pattern, pressed. **(H)**

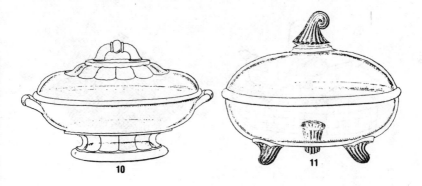

10　　　　　　　　　　11

12

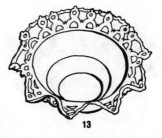

13

14

15

16. Covered entrée dish, M'Kee & Bros., Pittsburgh, 1890s, Dewey pattern, milk glass, pressed. **(F)**
17. Bonbon dish, T.B. Clark & Co., Honesdale, Pa., 1890s, St. George pattern, cut. **(F)**
18. Handled bonbon dish, T.B. Clark & Co., Honesdale, Pa., 1890s, Irving pattern, cut. **(F)**
19. Bonbon dish, T.B. Clark & Co., Honesdale, Pa., 1890s, Adonis pattern, cut. **(F)**
20. Bonbon dish, T.B. Clark & Co., Honesdale, Pa., 1890s, Dorrance pattern, cut. **(F)**
21. Nappy, maker unknown, 1890s, Corinthian pattern with 16-point hobstar, pressed. **(H)**

16

17

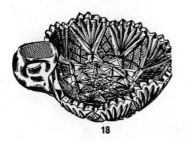

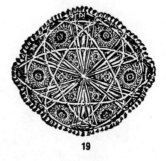

18

19

20

21

22. Covered dish, prob. The Central Glass Co., Wheeling, W. Va., 1890s, Swirl and Bottle pattern, pressed. **(G)**

23, 24. Bonbon dishes, prob. A. H. Heisey Glass Co., Newark, Ohio, late 1890s-1910, Diamond Lace or Hobstar pattern, pressed. **(H)**

25. Heart-shaped dish, maker unknown, early 1900s, Flying Bird with Strawberry pattern or Bluebird pattern, pressed. **(G)**

26. Banana boat, U.S. Glass Co., Pittsburgh, 1904, Colorado pattern, pressed. **(G)**

27. Olive dish, U.S. Glass Co., Pittsburgh, 1904, Victor pattern, pressed. **(H)**

28. Oblong dish, U.S. Glass Co., Pittsburgh, 1904, Victor pattern, pressed. **(H)**

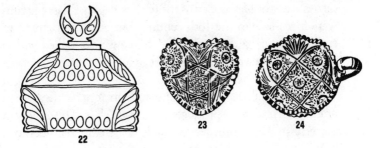

23   24

22

25

26

27   28

13. Egg Cups

Egg cups are charming reminders of the Victorian age. Although still made today, these small semi-ovoid-shaped dishes are as old-fashioned as the flask or flagon. Glass egg cups were first made in the 18th century, but became common in the home during the mid-1800s. Some were supplied with a domed top to keep the soft-boiled egg or eggs warm; a few were handled.

Egg cups, called by most manufacturers "egg holders," were among the first items produced in pressed glass. The variety of patterns in which they are available is therefore considerable. Shown are just a few of the many familiar patterns: Ashburton (no. 1), Texas Bull's Eye (no. 2), Huber (no. 3), New York or Honeycomb (no. 4), Eureka (no. 5), Sprig (no. 6), and Stedman (no. 7). The remaining examples are less common late-Victorian ware.

The most desirable mid-century pressed cups are those in opaque colors such as apple green, blue, and grease blue. The stems of opaque-colored egg cups are usually hexagonal, and the foot, circular. Later clear or clear colored cups, such as nos. 10 and 11, exhibit other stem and foot shapes.

Most egg cups are from 3″ to 3½″ high, and 5½″ high with cover. It is unusual today to find such covered pieces. As with other types of utilitarian ware, a top was easily broken or misplaced and rarely replaced.

1. Egg cup, maker unknown, New England or Pittsburgh area, 1840s, Ashburton pattern, pressed. **(G)**
2. Egg cup, prob. Bryce Bros., Pittsburgh, 1850-70, Bull's -Eye Variant or Texas Bull's-Eye pattern, pressed. **(G)**
3. Egg cup, New England Glass Co., E. Cambridge, Mass., 1860s, Huber pattern, pressed. **(G)**
4. Egg cup, M'Kee & Co., Pittsburgh, 1860s-70s, New York or Honeycomb pattern, pressed. **(H)**
5. Egg cup, M'Kee & Co., Pittsburgh, 1860s-70s, Eureka pattern, pressed. **(G)**
6. Egg cup, M'Kee & Co., Pittsburgh, 1860s-70s, Sprig or Ribbed Palm pattern, pressed. **(H)**
7. Egg cup, M'Kee & Co., Pittsburgh, 1860s-70s, Stedman pattern, pressed. **(G)**
8. Egg cup, King, Son & Co., Pittsburgh, 1870s, Jewel pattern, pressed. **(H)**
9. Egg cup, King, Son & Co., Pittsburgh, 1870s, No. 14 pattern, pressed. **(H)**
10. Egg cup, maker unknown, 1880s, Jewel Band pattern, pressed. **(H)**
11. Egg cup, maker unknown, 1890s, saucer foot, pressed. **(H)**

1 2 3 4

5 6 7 8

9 10 11

14. Goblets, and Cordial, Wine, and Champagne Glasses

GOBLETS—drinking vessels with a bowl supported by a stem—have long reflected elegance and sophistication. In Europe in the 16th and 17th centuries, these graceful objects were often covered with glass tops decorated with ornate finials and were seen only on the tables of the wealthiest families. In this country, goblets were made by the earliest glass companies. The free blown goblet on the following page (no. 1) is engraved with the date and place of making as well as with an intricate floral design; it is illustrative of the care and pride with which glassmakers executed these lovely stemmed drinking vessels.

While cut glass and blown designs continued to be made in the 19th century, especially in England (nos. 8,9), pressed-glass goblets were made by almost every American glass company in almost every pattern and style. Numerous variations on the goblet form were conceived in order to provide vessels of different sizes for use with a variety of beverages. The largest goblet usually stands approximately 5½" to 6¼" tall, while champagnes (nos. 13, 33, 34) stand 5", wines (nos. 23, 39) 3½" to 5", and cordials (nos. 11, 26, 42) a diminutive 2½" to 3½". Often two or three drinking vessels of the same bowl and stem shape can be found at a formal place setting: the largest holds water, the second wine, and the third may be used for an after dinner liqueur.

As goblets in over a thousand pressed-glass patterns were produced from the 1830s to the early years of this century, and these patterns are much reproduced today, it is important that the collector inspect each object carefully in order to discover distinguishing marks and design elements.

1. Goblet, New Bremen Glass Manufactory, Frederick, Md., 1792, engraved with ''G.F. Maverhoff'' and ''New Bremen, State of Maryland, Frederick County, Maryland, 1792,'' blown. **(A)**

1

2. Goblet, Boston & Sandwich Glass Co., 1830s, Single Vine Bellflower pattern, pressed. **(G)**
3. Goblet, maker unknown, Pittsburgh area or New England, 1840s-late 1870s, Ashburton pattern, pressed. **(F)**
4. Goblet, prob. Boston & Sandwich Glass Co., c. 1850, Flaring Grooved Bigler pattern or Worchester pattern, pressed. **(G)**
5. Goblet, New England Glass Co., E. Cambridge, Mass., 1850s, Bull's Eye pattern, pressed. **(F)**
6. Goblet, prob. The Union Glass Co., Somerville, Mass., 1850s-80s, Buckle pattern, pressed. **(H)**
7. Goblet, maker unknown, Pittsburgh area or New England, 1850s-80s, Flute pattern, pressed. **(G)**
8. Goblet, Richardson, Stourbridge, England, 1851, crystal, etched design, blown. **(F)**
9. Goblet, Richardson, Stourbridge, England, 1851, crystal, cut. **(F)**
10. Goblet, Boston & Sandwich Glass Co., c. 1860, Lincoln Drape with Tassel pattern, pressed. **(F)**

2

3

4

5

6

7

8

9

10

11. Cordial glass, maker unknown, Pittsburgh area, 1860s, Tulip pattern, pressed. **(F)**
12. Goblet, M'Kee & Bros., Pittsburgh, 1864, Eugenie pattern, pressed. **(F)**
13. Champagne glass, M'Kee & Bros., Pittsburgh, 1868, Stedman pattern, pressed. **(F)**
14. Goblet, prob. Richards and Hartley Flint Glass Co., Pittsburgh, 1870s, Loop and Dart pattern with diamond ornaments, pressed. **(G)**
15. Goblet, Central Glass Co., Wheeling, W. Va., 1870s, Cabbage Rose pattern, pressed. **(F)**

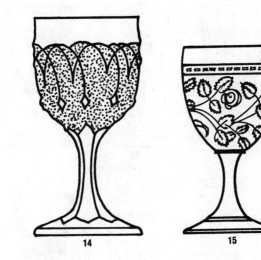

16. Goblet, King, Son & Co., Pittsburgh, 1870s, No. 127 pattern, pressed. **(G)**
17. Goblet, King, Son & Co., Pittsburgh, 1870s, Mitchell A ware, pressed. **(H)**
18. Goblet, King, Son & Co., Pittsburgh, 1870s, Mitchell B ware, pressed. **(H)**
19. Goblet, King, Son & Co., Pittsburgh, 1870s, Mitchell C ware, pressed. **(H)**
20. Goblet, King, Son & Co., Pittsburgh, 1870s, Mitchell D ware, pressed. **(H)**
21. Goblet, King, Son & Co., Pittsburgh, 1870s, Eng. No. 34 pattern, pressed. **(H)**
22. Goblet, King, Son & Co., Pittsburgh, 1870s, Girard pattern, pressed. **(G)**
23. Wine glass, King, Son & Co., Pittsburgh, 1870s, Berlin pattern, pressed. **(H)**
24. Goblet, King, Son & Co., Pittsburgh, 1870s, Jewel pattern, pressed. **(G)**

16

17

18

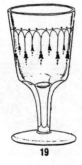
19

20

21

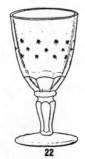
22

23

24

25. Wine glass, King, Son & Co., Pittsburgh, 1870s, Greek pattern, pressed. **(H)**
26. Cordial glass, King, Son & Co., Pittsburgh, 1870s, Havana pattern, pressed. **(H)**
27. Goblet, King, Son & Co., Pittsburgh, 1870s, Lattice pattern, pressed. **(G)**
28. Goblet, King, Son & Co., Pittsburgh, 1870s, Huber pattern, pressed. **(G)**
29. Goblet, King, Son & Co., Pittsburgh, 1870s, Pearl pattern, pressed. **(H)**
30. Goblet, King, Son & Co., Pittsburgh, 1870s, Gothic ware, pressed. **(G)**
31. Goblet, M'Kee & Bros., Pittsburgh, 1870s, Cincinnati pattern, pressed. **(H)**
32. Goblet, Bellaire Goblet Co., Findlay, Ohio, 1870s, 101 pattern, pressed. **(G)**
33. Champagne glass, maker unknown, 1880s-90s, Pillar Cuts pattern, cut. **(H)**

25

26

27

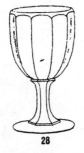
28

29

30

31

32

33

34. Champagne glass, maker unknown, 1880s-90s, Strawberry Diamonds pattern, pressed. **(H)**

35. Sherry glass, maker unknown, 1880s-90s, Middlesex variant, cut. **(G)**

36. Goblet, prob. Adams & Co., Pittsburgh, 1880s-90s, Ruby Thumbprint or King's Crown pattern, pressed. **(G)**

37. Goblet, prob. Doyle & Co., Pittsburgh, c. 1885, Triple Triangle pattern, pressed. **(G)**

38. Goblet, M'Kee & Bros., Pittsburgh, 1890s, Beaded Tulip pattern, pressed. **(G)**

39. Wine glass, maker unknown, c. 1895, Fancy pattern, pressed. **(H)**

34

35

36

37

38

39

40. Goblet, maker unknown, c. 1895, pressed. **(H)**
41. Goblet, Dalzell, Gillmore & Leighton, Findlay, Ohio, late 1890s, Priscilla pattern, pressed. **(G)**
42. Cordial glass, T.B. Clark & Co., Honesdale, Pa., 1896, Winola pattern, cut. **(G)**
43. Goblet, U.S. Glass Co., Pittsburgh, 1900s, Virginia pattern, pressed. **(H)**
44. Goblet, U.S. Glass Co., Pittsburgh, 1900s, Pennsylvania pattern, pressed. **(H)**
45. Goblet, Anchor Hocking Glass Co., Lancaster, Ohio, 1939-60s, Royal Ruby pattern, pressed. **(H)**

15. Household Articles

IMPROVEMENTS in the manufacture of glass during the 1800s brought about the common use of the material in various ways—as knobs and tiebacks, bells, match holders and toothpick holders, ashtrays and spittoons, ink stands, and assorted feeding devices for infants and pets. Although primarily utilitarian in design and use, these glass objects are now highly collectible.

Knobs and tiebacks were among the first items to be made by manufacturers of pressed glass in the 1820s and '30s and continued to be made through the early 1900s. The most beautiful and valuable of these objects are the iridescent white tiebacks, such as no. 1, made by the Boston & Sandwich Glass Co. as well as other New England and western Pennsylvania glasshouses.

Throughout the 19th century the major makers of pressed glass also turned out mold-blown objects and often free-blown as well. Glass hats, used as toothpick holders, were first produced as mold-blown items, but later were also pressed (no. 19). Globes and smoke bells (no. 3) were mold blown or free blown. Inkstands were both pressed and mold blown; globes for goldfish, hung in many Victorian parlors from chains, were free blown, as were dinner bells used at the table. Manufacturers of cut glass supplied such useful and decorative objects as knife rests.

Glass also found a handy use in the kitchen as butter prints (no. 8), sugar sifters (nos. 21 and 22), lemon and orange juice extractors (no. 23), and even as rolling pins.

1. Tieback, prob. Boston & Sandwich Glass Co., 1835-45, opalescent, pressed. **(G)**
2. Seed box, M'Kee & Bros., Pittsburgh, 1864, pressed. **(H)**
3. Smoke bell, M'Kee & Bros., Pittsburgh, 1864, mold blown. **(G)**
4. Bird fountain, M'Kee & Bros., Pittsburgh, 1864, pressed. **(H)**
5. Well inkstand, M'Kee & Bros., Pittsburgh, 1864, pressed. **(H)**
6. Spittoon, M'Kee & Bros., Pittsburgh, 1864, pressed. **(H)**

1

2

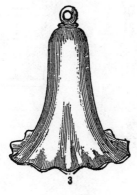

3

4

5

6

7. Bird bath, M'Kee & Bros., Pittsburgh, 1868, pressed. **(H)**
8. Butter print, M'Kee & Bros., Pittsburgh, 1868, pressed. **(G)**
9. Medicine glass, King, Son & Co., Pittsburgh, 1870s, pressed. **(H)**
10. Inkstand, King, Son & Co., Pittsburgh, 1870s, No. 13 pattern, pressed. **(H)**
11. Hand bell, maker unknown, late 1800s, blown. **(G)**
12. Match holder, maker unknown, late 1800s, oaken bucket form, pressed. **(G)**
13. Knife rest, maker unknown, late 1800s, cut. **(H)**
14. Inkstand, Whitall, Tatum & Co., Millville, N.J., 1880s, Fluted Pyramid form, mold blown. **(H)**

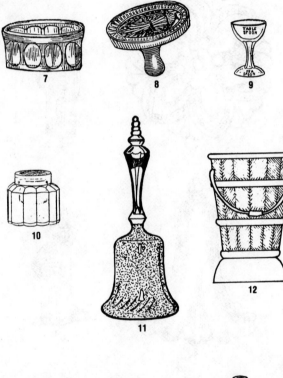

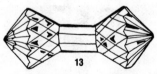

15. Hanging fish globes, Whitall, Tatum & Co., Millville, N.J., 1880s, mold blown. **(G)**

16. Mucilage holder, Whitall, Tatum & Co., Millville, N.J., 1880s, cone form, mold blown. **(H)**

17. Match holder, maker unknown, 1880s-90s, clock form, pressed. **(F)**

18. Toothpick holder, maker unknown, c. 1895, Ceylon design, pressed. **(H)**

19. Toothpick holder, maker unknown, c. 1895, hat-shaped, pressed. **(G)**

20. Toothpick holder, maker unknown, c. 1895, pressed. **(H)**

15

16

17

18   19   20

21. Sugar sifter, maker unknown, c. 1895, quilt pattern, pressed. **(H)**
22. Sugar sifter, maker unknown, c. 1895, opalescent, pressed. **(G)**
23. Lemon extractor, maker unknown, c. 1895, Manny glass, pressed. **(H)**
24. Bell, T.B. Clark & Co., Honesdale, Pa., 1896, Jewel pattern, cut.. **(F)**
25. Hatpin holder, maker unknown, early 1900s, mold blown. **(H)**
26. Toothpick holder, U.S. Glass Co., Pittsburgh, 1900s, Oregon pattern, pressed. **(H)**
27. Ash tray, Jeannette Glass Co., Jeannette, Pa., 1936-46, Windsor Diamond pattern, pressed. **(G)**

21   22   23

26

25   27

24

16. Jars

J ARS for storing various kinds of food have been made of glass for hundreds of years. The most common were made for jam, pickles, or mustard; the most elaborate were showpieces for commercial display. Most jars produced in the 1800s are pressed rather than mold blown or free blown. All have covers or lids and most are cylindrical in shape.

Makers of pressed glass often included a jar for jam, pickles, or mustard in their pattern sets. Representative of this type of container are such pattern pieces as Westward-Ho (no. 7), Barley (no. 8), Hexagonal Block (no. 9), and Hobnail (no. 11). Jars were also sometimes called "casters" and placed in a handled metal carrier of some sort. A caster set might include three or four jars or merely stand alone, as in no. 13. The late-Victorian household made use of more and more specialty containers. Some of the more unusual jars were intended for biscuits or crackers (nos. 15 and 18) or to hold straws (no. 17).

Jars for use by pharmacists include showpieces (such as no. 4) and urns (no. 5). Various types of medicinal and cosmetic preparations were routinely packaged in plain jars. Tooth powder or cold cream came in the small container (no. 10), made by Whitall, Tatum & Co. in the 1880s. Pomade or other ointments might be packaged in the larger jar (no. 6), also made by Whitall, Tatum. Containers such as these are relatively easy to find today and usually include a lettered lid or label.

1. Honey jar, Waterford, Ireland, 1700s, Raised Diamond pattern, cut. **(D)**
2. Oval pickle jar, King, Son & Co., Pittsburgh, 1870s, No. 13 pattern, pressed. **(H)**
3. Round pickle jar, King, Son & Co., 1870s, No. 13 pattern, pressed. **(H)**
4. Show jar, Whitall, Tatum & Co., Millville, N.J., 1880, blown. **(E)**

5. Counter urn, Whitall, Tatum & Co., Millville, N.J., 1880, blown. **(F)**

6. Round pomade jar, Whitall, Tatum & Co., Millville, N.J., 1880, blown. **(H)**

7. Marmalade jar, Gillinder & Sons, Philadelphia, 1880s, Westward-Ho pattern, pressed. **(E)**

8. Jam jar, Campbell, Jones & Co., Pittsburgh, 1880s, Barley pattern, pressed. **(F)**

9. Pickle jar, maker unknown, Midwest, 1880s, Hexagonal block pattern, pressed. **(H)**
10. Ointment or tooth powder jar, Whitall, Tatum & Co., Millville, N.J., 1880, pressed. **(H)**
11. Handled mustard jar, prob. New Brighton Glass Co., New Brighton, Pa. or Hobbs, Brockunier & Co., Wheeling, W. Va., 1880s-90s, Hobnail pattern, pressed. **(G)**
12. Mustard pot, prob. Hobbs, Brockunier & Co., Wheeling, W. Va., c. 1895, Dewdrop pattern, pressed. **(G)**
13. Pickle caster, prob. M'Kee & Bros., Pittsburgh, c. 1895, Nellie pattern, pressed. **(G)**
14. Pickle jar, maker unknown, prob. Pittsburgh area, c. 1880s-90s, Pillar pattern, pressed. **(G)**

9

10

11

12

13

14

15. Biscuit jar, U.S. Glass Co., Pittsburgh, 1900s, New Hampshire pattern, pressed. **(G)**
16. Pickle jar, U.S. Glass Co., Pittsburgh, 1900s, Pennsylvania pattern, pressed. **(G)**
17. Straw jar, U.S. Glass Co., Pittsburgh, 1900s, Pennsylvania pattern, pressed. **(G)**
18. Cracker jar, U.S. Glass Co., Pittsburgh, 1900s, pressed. **(G)**

16

15

17

18

17. Jugs and Cruets

GLASS jugs and flagons for syrup and liquor have been common items in homes and public dining places since the 18th century. Cruets for liquid condiments were also used in the 1700s, but became most popular in the Victorian era. Several cruet bottles were most likely to be arranged in a glass or metal stand.

Many early jugs and flagons were mold blown with applied blown handles. Completely pressed jugs, however, were produced by the mid-1800s. What many blown or pressed varieties have in common are a globular shape, a wide mouth, perhaps a pouring lid or spout, a stopper or lid, and a loop handle. Cruets follow the same general lines. They are often inscribed with the name of the condiment—oil, vinegar, lemon juice, garlic juice, etc.

Glass jugs are less common than those made of pottery and were in their time more expensive. They are also more highly decorated. The most fanciful of the designs (nos. 1 and 2) appear on English jugs. Most elaborate of all are the blown and cut-glass cruets and jugs (nos. 10-12) made in the 1890s and early 1900s. Sterling silver tops were sometimes supplied for the fanciest types, such as no. 12 made by T. B. Clark & Co.

1. Jug, Richardson, Stourbridge, England, 1851, crystal, mold blown. **(F)**
2. Claret jug, maker unknown, English, 1851, engraved, mold blown. **(F)**
3. Syrup jug, maker unknown, New England or Pittsburgh area, 1850s-60s, fluted, mold blown. **(G)**
4. Syrup jug, M'Kee & Bros., Pittsburgh, 1860s, Pillar pattern, pressed. **(H)**
5. Cruet, O'Hara Glass Co., Ltd., Pittsburgh, 1880s, Sawtooth and Star pattern, pressed. **(H)**

1

2

3

4

5

6. Syrup jug, maker unknown, Pittsburgh area, 1880s, Lion pattern, pressed. **(F)**
7. Cruet, Bryce, M'Kee & Co., Pittsburgh, 1890s, Berkeley pattern, pressed. **(G)**
8. Syrup jug, maker unknown, c. 1895, pressed. **(H)**
9. Vinegar cruet, maker unknown, Pittsburgh area, c. 1895, Jeanette pattern, pressed. **(H)**
10. Vinegar cruet, maker unknown, c. 1895, cut. **(G)**

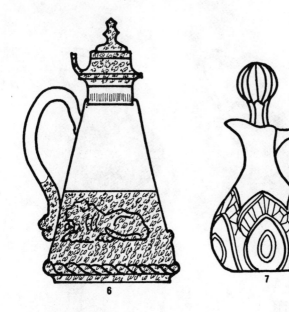

6

7

8

9

10

11. Whiskey jug, T.B. Clark & Co., Honesdale, Pa., 1890s, Venus pattern, cut. **(D)**

12. Claret jug, T.B. Clark & Co., Honesdale, Pa., 1890s, Arbutus pattern with No. 202 A sterling top, 1890s, cut. **(D)**

13. Molasses jug, U.S. Glass Co., Pittsburgh, 1900s, Georgia pattern, pressed. **(H)**

14. Syrup jug, U.S. Glass Co., Pittsburgh, 1900s, Pennsylvania pattern, pressed. **(H)**

11   12

13   14

18. Lamps

PRODUCED in a wide variety of designs and styles in the mid-
and late 19th century, fluid burning lamps are among the most
valuable and sought-after glass collectibles. Hand or chamber
lamps are one of the most commonly collected types of lamps.
Used to illuminate the bedchamber just before retiring, they
were often lit in one room of the house and carried to the
bedroom; small and easily portable, they are usually handled
(nos. 1,2,13). Night lamps can be distinguished from chamber
lamps as they are handleless and were permanent fixtures on
the bureau or bedside table (no. 14). Larger lamps for use in the
main rooms of the house were often made in two parts: the
base was pressed (no. 8) and the font was mold blown. The two
elements were often joined with thin bands or knobs of blown
glass known as knops. In the late 1800s both base and font
were pressed in one piece. Lanterns, such as the globe lantern
made by M'Kee (no. 5), were commonly used to light the way
out of doors, both by those at home and those who were forced
to travel at night.

As the century progressed, more elaborately designed and
colored lamps began to be created. The two lamps attributed
to Boston & Sandwich Glass Co. (nos. 10,11) have ornate
overlaid fonts. The lamp made by Ripley & Co. (no. 12) is
known as a bride's or marriage lamp and has a white opaque
base, clambroth match holder, and opaque blue fonts. The
most decorative lamps were created in the Tiffany Studios
around the turn of the century (no. 18). With their intricate
stained-glass shades, Tiffany lamps are much admired and
reproduced today, and originals are among the most precious
of all glass objects.

1. Chamber lamp, maker unknown, mid-1800s, paneled oval designs, brass fluid burner, pressed. **(E)**
2. Oil lamp, M'Kee & Bros., Pittsburgh, 1864, ribbed design, blown with applied molded handle. **(F)**
3. Oil lamp, M'Kee & Bros., Pittsburgh, 1864, Prism pattern, pressed and blown. **(F)**
4. Oil lamp, M'Kee & Bros., Pittsburgh, 1864, Turnip pattern, pressed and blown. **(F)**
5. Globe lantern, M'Kee & Bros., Pittsburgh, 1864, blown globe. **(H)**

1

2

3

4

5

6. Oil lamp, M'Kee & Bros., Pittsburgh, 1864, Tulip pattern, pressed and blown. **(F)**
7. Oil lamp, M'Kee & Bros., Pittsburgh, 1864, Vine pattern, pressed and blown. **(F)**
8. Lamp base, M'Kee & Bros., Pittsburgh, 1864, pressed. **(H)**
9. Oil lamp, M'Kee & Bros., Pittsburgh, 1864, ribbed design, pressed and blown. **(F)**
10. Oil lamp, Boston & Sandwich Glass Co., 1865-70, opaque white base, pressed and blown. **(E)**
11. Oil lamp, prob. Boston & Sandwich Glass Co., 1865-70, overlay design, pressed and blown. **(D)**

12. Lamp, Ripley & Co., Pittsburgh, patented 1870, opaque blue blown fonts, white opaque pressed base and match holder. **(B)**

13. Oil lamp, M'Kee & Bros., Pittsburgh, 1868, Ring pattern, blown and molded with applied molded handle. **(F)**

14. Night lamp, M'Kee & Bros., Pittsburgh, 1868, blown. **(G)**

15. Oil lamp, M'Kee & Bros., Pittsburgh, 1868, Argus pattern, blown and pressed. **(F)**

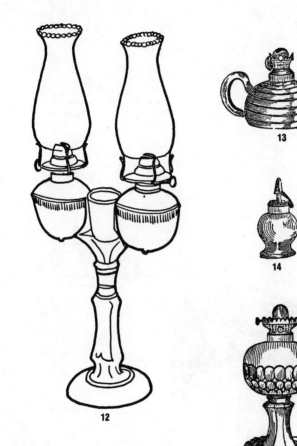

16. Lamp, King, Son & Co., Pittsburgh, 1870, pressed and blown. **(F)**
17. Lamp, King, Son & Co., Pittsburgh, 1870, fluted design, pressed and blown. **(F)**
18. Lamp, Tiffany Studios, New York, N.Y., c. 1900-05, Wistaria design, stained glass. **(A)**

16

17

18

19. Miniatures

MINIATURE glass objects, toy versions of larger pieces, were widely produced in the 19th century for the delight of children of all ages. The objects were free blown, mold blown, and pressed in a great variety of forms and designs. Some of the pieces were intended for use with dollhouse furniture; others, such as the ABC and 1-2-3 plates (nos. 7 and 8, respectively) are pedagogical devices to facilitate learning the basics.

Among the earliest producers of miniature objects were the Boston & Sandwich Glass Co. and various Pittsburgh-area firms such as M'Kee & Bros. The pressed-glass toy tumbler, sad iron, and candlestick (nos. 1-3) were produced by M'Kee in clear glass and various colors. These forms bring extraordinarily high prices today since few have survived the wear and tear of childhood. The Boston & Sandwich Glass Co. is also known to have made lacy glass toy objects in the 1830s and '40s.

Later Victorian miniatures are more highly decorative than earlier pieces and reflect the general trend toward elaboration in design which marked standard-sized objects from the 1870s to the early 1900s. The toy mug with a bird design (no. 9) is representative of the type of miniature produced during this period. Like the earlier pieces, the high-Victorian miniatures fetch a very high price today.

1. Tumbler, M'Kee & Bros., Pittsburgh, late 1860s, pressed. **(G)**
2. Sad iron, M'Kee & Bros., Pittsburgh, late 1860s, pressed. **(F)**
3. Candlestick, M'Kee & Bros., Pittsburgh, late 1860s, pressed. **(F)**
4. Mug, King, Son & Co., Pittsburgh, 1870s, Vine pattern, pressed. **(G)**
5. Toy set, King, Son & Co., Pittsburgh, 1870s, frosted No. 13 pattern, pressed. **(F)**
6. Toy set, King, Son & Co., Pittsburgh, 1870s, No. 13 pattern, pressed. **(F)**
7. ABC plate, King, Son & Co., Pittsburgh, 1870s, pressed. **(G)**
8. 1, 2, 3 plate, King, Son & Co., Pittsburgh, 1870s, pressed. **(G)**
9. Mug, Bryce, M'Kee & Co., Pittsburgh, late 1800s, bird design, pressed. **(G)**
10. Toothpick holder, maker unknown, c. 1895, toy bucket, pressed. **(G)**

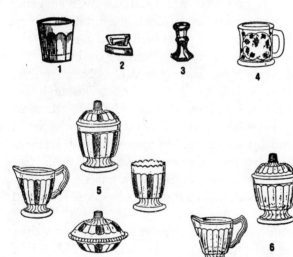

20. Mugs

Usually made of pottery and used for hot drinks such as coffee, tea, and hot chocolate, mugs are among the most common drinking vessels. Capacious and sturdy, mugs are used on an everyday basis, while more delicate cups with saucers are used at formal occasions (see section 10). Mugs made of glass are usually of much the same design as those of pottery—flat-based with applied handles—but were not commonly made and are now valuable and difficult to find. The blown-glass mug attributed to the New England Glass Co. (no. 1 on the following page) is an early example and is particularly unusual; it is footed and engraved with the initials "WW" and the date "July 18, 1842." Unlike this one, later mugs were rarely personalized. Pressed-glass mugs are decorated with both geometric (nos. 9, 10) and naturalistic patterns (nos. 4,7,12) and were produced in both the Pittsburgh area and New England.

Steins or pilseners were brought to this country by German immigrants in the middle of the last century. The habit of using these heavy glass mugs, with a capacity of one or two pints of liquid, to contain beer became popular in this country, and beer mugs were soon produced by American pressed-glass companies (nos. 3,5,6). The use of glass beer mugs has become a custom in America and has made the beer mug the most common type of glass mug found today.

1. Mug, prob. New England Glass Co., E. Cambridge, Mass., 1842, engraved and threaded, blown. **(G)**
2. Mug, Atterbury & Co., Pittsburgh, c. 1870, Medallion pattern, pressed. **(H)**
3. Beer mug, King, Son & Co., Pittsburgh, 1870s, pressed. **(H)**
4. Mug, King, Son & Co., Pittsburgh, 1870s, Vine pattern, pressed. **(H)**
5. Beer mug, King, Son & Co., Pittsburgh, 1870s, Pilsener style, pressed. **(H)**
6. Beer mug, King, Son & Co., Pittsburgh, 1870s, stuck handle style, pressed. **(H)**

1

2

3

5

4

6

7. Mug, prob. Bryce, M'Kee & Co., Pittsburgh, 1880s, Wheat and Barley pattern, pressed. **(H)**
8. Mug, prob. The Brilliant Glass Works, Brilliant, Ohio, c. 1885, Cut Log or Cat's Eye and Block pattern, pressed. **(H)**
9. Mug, maker unknown, Pittsburgh area or New England, late 1800s, Ashburton pattern, pressed. **(G)**
10. Mug, maker unknown, Pittsburgh area or New England, late 1800s-early 1900s, Beaded Circle pattern, pressed. **(G)**
11. Mug, U.S. Glass Co., Pittsburgh, 1900s, Wyoming pattern, pressed. **(H)**
12. Mug, U.S. Glass Co., Pittsburgh, 1900s, Georgia pattern, pressed. **(H)**

7

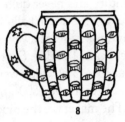
8

9

10

11

12

21. Novelties

GLASSMAKERS have always delighted in turning out fanciful objects which exhibit a workman's skill. The art of free blowing glass was widely practiced in the 1700s and early 19th century and never quite died away in later years. Glass balls, often called "witch balls" because they were supposedly hung in windows to scare away evil spirits, are beautifully formed objects (nos. 1 and 2). So, too, are receptacles in the form of top hats (nos. 3-6) which were free blown and mold blown at first and later in the 1800s, pressed.

The makers of the models used to create molds for pressing glass are the individuals responsible for the "artistry" of pressed novelties. Some of the most common late-19th-century forms they created are the high-button shoe, the roller skate (no. 10), the boot, and the slipper (no. 13). There are hundreds of such objects, and many of them could be used as toothpick or match holders.

Other novelty items are merely knick-knacks made to grace the shelves of a Victorian what-not. Representative of this type are the kettle (no. 8), basket (no. 9), washboard and tub (nos. 14, 15), and cradle (no. 16). As pressed-glass objects, they were not very expensive when made. Because tens of thousands of such items were pressed, their price today is still reasonable.

1. Glass ball and bowl, maker unknown, mid 1800s, blown. **(E)**
2. Witch ball, maker unknown, mid 1800s, swirl decoration, blown. **(D)**
3. Top hat, maker unknown, mid 1800s, spattered decoration, mold blown. **(G)**
4. Top hat, maker unknown, mid 1800s, threaded, mold blown. **(G)**
5. Top hat, maker unknown, mid 1800s, clear, blown. **(G)**
6. Top hat, maker unknown, 1880s-90s, Daisy and Button pattern, pressed. **(H)**
7. Cradle, The New England Glass Co., E. Cambridge, Mass., c. 1886, variant Russian pattern, cut. **(F)**

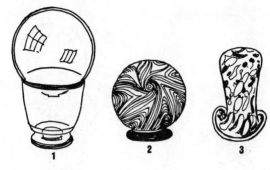

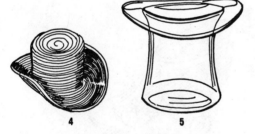

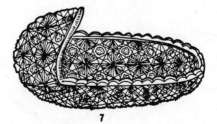

8. Kettle matchholder, maker unknown, 1880s-90s, Diamond Thumbprint pattern, pressed. **(G)**

9. Basket, maker unknown, 1880s-90s, basket design with rope handle, pressed. **(G)**

10. Roller skate toothpick holder, U.S. Glass Co., Pittsburgh, 1890s, Daisy and Button pattern, pressed. **(H)**

11, 12. Boots, maker unknown, late 1800s, translucent white (left), cranberry swirled (right), mold blown. **(H)**

13. Slipper matchholder, maker unknown, late 1800s, pressed. **(H)**

14. Washboard, U.S. Glass Co., Pittsburgh, 1890s, pressed. **(H)**

15. Washtub, U.S. Glass Co., Pittsburgh, 1890s, pressed. **(H)**

16. Cradle, maker unknown, late 1800s, pressed. **(H)**

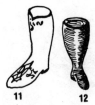

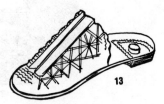

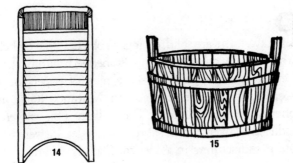

14

15

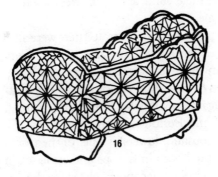

16

22. Paperweights

PAPERWEIGHTS are solid, clear glass objects often embellished with intricate colored designs embedded within the glass. They are usually used to secure piles of papers on desks. Paperweights were first made in Murano, Italy, and in Bohemia in the mid-1840s. The brilliant objects were soon produced all over Europe, and within a decade were made by glass manufacturers in America. Slightly magnified by the dome of glass above them, the designs inside appear different from various angles, and, as they are free blown, no two are exactly the same.

The nine paperweights illustrated here were all made in New England and New Jersey. The cameo (no. 1) contains a profile of Benjamin Franklin; cameo paperweights are often filled with profiles of other famous (or simply beautiful) faces. The poinsettia against a white latticinio background (no. 2) was a specialty of the Boston & Sandwich Glass Co. Millefiori (literally a thousand flowers) is one of the most common of designs executed both in Europe and America (no. 4) and is made by bunching tiny canes of glass into the form of florets. This example, also made by the Boston & Sandwich Glass Co., has facets cut into the dome which serve to frame the flowers. Whitall, Tatum & Co., of Millville, N.J., produced several innovative designs, including the famous Millville rose (no. 9) and the rare tulip (no. 7). Abstract decorations such as the red, white, and blue spiral twists produced by Pairpoint Corp., are more unusual than illusionistic paperweight designs. As very few paperweights are marked with dates, it is important for the collector of these popular objects to remember that the value of the piece is based on the relative rarity or commonness of the decorative elements.

1. Paperweight, New England Glass Co., E. Cambridge, Mass., mid 1800s, Ben Franklin cameo, blown. **(C)**
2. Paperweight, Boston & Sandwich Glass Co., mid 1800s, poinsettia, blown. **(D)**
3. Paperweight, Mt. Washington Glass Works, New Bedford, Mass., 1850s-80s, strawberries, blown. **(C)**
4. Paperweight, Boston & Sandwich Glass Co., 1850s-80s, Millefiori, blown. **(C)**
5. Paperweight, New England Glass Co., E. Cambridge, Mass., c. 1860, pear on base, blown. **(D)**
6. Paperweight, New England Glass Co., E. Cambridge, Mass., c. 1860, apples and pears, blown. **(D)**
7. Paperweight, Whitall, Tatum & Co., Millville, N.J., early 1900s, tulip in globe, blown. **(D)**
8. Paperweight, Pairpoint Corp., New Bedford, Mass., early 1900s, Red, white, and blue spiral twists, blown. **(E)**
9. Paperweight, Whitall, Tatum & Co., Millville, N.J., c. 1905-12, Millville rose, blown. **(D)**

23. Perfume and Cologne Bottles

Tiny bottles holding sweet-smelling colognes and perfumes have long graced dressing tables and bureaus in boudoirs and bedrooms. America's earliest glass manufacturers produced these luxury items (as well as snuff bottles, see no. 1 on the following page), and the free blown amethyst-colored scent bottle of the Stiegel type (no. 2) is certainly one of the loveliest of its kind. The Boston & Sandwich Glass Company produced many fine mold blown toilet bottles in the mid-19th century (no. 4), and it is a collector's dream to discover one of these, especially if its stopper is intact and original (nos. 5, 6).

Colognes and perfumes were most often purchased from the local druggist; it was he who packaged the dulcet liquid and he who chose the bottles in which it was held. Most often druggists preferred the inexpensive styles available from a company like Whitall, Tatum (nos. 12-28). This company supplied a wide variety of cut and mold-blown perfume and cologne bottles; its cologne bottles were generally somewhat larger than the perfume bottles and often had tapered necks (nos. 16, 17, 18). Alternatively, the cheaper bottle from the druggist might have been taken home and discarded in favor of something more elegant like the cut-glass colognes made by T.B. Clark & Co. (nos. 30, 31, 32). These were permanent fixtures on the dressing table and were reused indefinitely. Fancy perfume bottles are often found with silver or metal tops. Perfume and cologne bottles were decorated in all manner of styles, although, more often than not, geometric patterns were preferred to more naturalistic designs.

1. Snuff bottle, prob. Stiegel, Pennsylvania, 1770s, blown. **(D)**
2. Scent bottle, Stiegel-type, Pennsylvania, 1774-90, ogival design on shoulder, vertical fluting below, blown. **(E)**
3. Cologne bottle, English, late 1700s, white Bristol glass with deep-blue decoration, blown. **(F)**
4. Scent bottle, Boston & Sandwich Glass Co., 1820s-30s, geometric pattern, mold blown. **(F)**
5. Perfume bottle, Boston & Sandwich Glass Co., 1830s-40s, blown-three-mold. **(F)**
6. Perfume bottle, Boston & Sandwich Glass Co., 1830s-40s, blown-three-mold. **(F)**

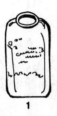

1

2

3

4

5

6

7. Cologne bottle, maker unknown, mid 1800s, block-cut. **(E)**
8. Snuff bottle, maker unknown, mid 1800s, cameo glass, blown. **(E)**
9. Cologne bottle, maker unknown, 1870s, Trilby pattern, pressed. **(F)**
10. Perfume vial, prob. New England Glass Co., E. Cambridge, Mass., 1870s-80s, enameled glass, blown. **(F)**
11. Perfume vial, maker unknown, 1880s, cut. **(F)**
12. Perfume bottle, Whitall, Tatum & Co., Millville, N.J., 1880s, Extra Heavy Square design, mold blown. **(H)**
13. Perfume bottle, Whitall, Tatum & Co., Millville, N.J., 1880s, Fancy Square Ball-Neck Panel design, mold blown. **(G)**

7

8

9

10

11

12

13

14. Perfume bottle, Whitall, Tatum & Co., Millville, N.J., 1880s, Diamond Blake design, cut. **(G)**
15. Perfume bottle, Whitall, Tatum & Co., Millville, N.J., 1880s, Round Brilliant design, cut. **(F)**
16. Cologne bottle, Whitall, Tatum & Co., Millville, N.J., 1880s, Cone pattern, painted, mold blown. **(G)**
17. Cologne bottle, Whitall, Tatum & Co., Millville, N.J., 1880s, Diamond pattern, cut. **(F)**
18. Cologne bottle, Whitall, Tatum & Co., Millville, N.J., 1880s, Caswell design, mold blown. **(H)**
19. Perfume bottle, Whitall, Tatum & Co., Millville, N.J., 1880s, Fancy Square Panel design, mold blown. **(H)**
20. Perfume bottle, Whitall, Tatum & Co., Millville, N.J., 1880s, jug handle, mold blown. **(G)**

14

15

16

17

18

19

20

21. Perfume bottle, Whitall, Tatum & Co., Millville, N.J., 1880s, Plain Lubin design, mold blown. **(H)**
22. Perfume bottle, Whitall, Tatum & Co., Millville, N.J., 1880s, Cone Heavy Base design, mold blown. **(G)**
23. Perfume bottle, Whitall, Tatum & Co., Millville, N.J., 1880s, Oval Brilliant design, cut. **(F)**
24. Cologne bottle, Whitall, Tatum & Co., Millville, N.J., 1880s, Champagne pattern, mold blown. **(G)**
25. Perfume bottle, Whitall, Tatum & Co., Millville, N.J., 1880s, fluted foot, mold blown. **(G)**
26. Perfume bottle, Whitall, Tatum & Co., Millville, N.J., 1880s, Fancy Rhombus design, mold blown. **(F)**
27. Phenix Atomizer, Whitall, Tatum & Co., Millville, N.J., 1880s, Bell style, pressed and mold blown. **(G)**

21   22   23   24

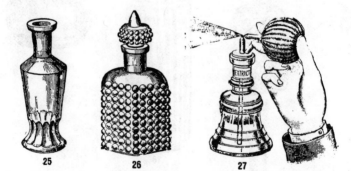

25   26   27

28. Cologne bottle, Whitall, Tatum & Co., Millville, N.J., 1880s, cut stopper, gilded, mold blown. **(F)**
29. Puff box, O'Hara Glass Co., Ltd., Pittsburgh, Block and Panel pattern, 1880s, pressed. **(G)**
30. Globe cologne bottle, T.B. Clark & Co., Honesdale, Pa., 1896, Venus pattern, cut. **(F)**
31. Cologne bottle, T.B. Clark & Co., Honesdale, Pa., 1896, Jewel pattern, cut. **(F)**
32. Cologne bottle, T.B. Clark & Co., Honesdale, Pa., 1896, St. George pattern, cut. **(F)**

28

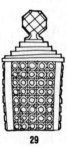

29

30

31

32

24. Pitchers and Tankards

Pitchers were among the earliest utensils made by man, and are still among the most useful. Archaeological expeditions have uncovered the ancient form in a variety of materials, ranging from the humblest clay pieces to beautiful examples wrought in pure gold. The ubiquitous pitcher was one of the first pieces made by American glassmakers in the 17th and 18th centuries. Early examples were hand blown, generally with applied decorative motifs (no. 1). Later forms were mold blown, a process in which the glass artisan first blew into a small mold, then removed the piece to enlarge it. This technique was widely used in England as well (nos. 3 and 4).

When pressed glass became readily available in the 1820s and '30s, pitchers became plentiful in every home, as the assembly-line technique made them much cheaper. Often they were available in sets (nos. 18 and 19), with matching tumblers and waste bowl, along with a complementary tray, sometimes made of metal.

Elaborately cut pitchers, such as T. B. Clark & Company's Adonis pattern (no. 23), were extremely popular during the late 19th century. Their ornate, deeply-cut patterns reflected the opulence of the late-Victorian age, though they were certainly not practical pieces. Even empty, such pitchers tended to be very heavy and unwieldy.

1. Pitcher, maker unknown, South Jersey, early 1800s, applied threads and trails, blown. **(D)**
2. Pitcher, maker unknown, American or English, 1840s-50s, pressed. **(G)**
3. Pitcher, Richardson, Stourbridge, England, 1850s, mold blown. **(F)**
4. Water pitcher, maker unknown, English, 1850s, mold blown. **(F)**
5. Pitcher, M'Kee & Bros., Pittsburgh, 1860s, Bellflower or Ribbed Leaf, pressed. **(E)**

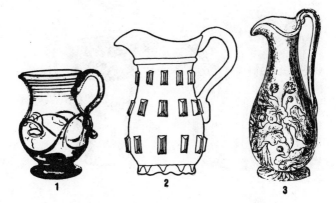

6. Pitcher, M'Kee & Bros., Pittsburgh, 1860s, Ribbed pattern, pressed. **(F)**
7. Pitcher, Adams & Co., Pittsburgh, 1870s, Wild Flower pattern, pressed. **(F)**
8. Pitcher, M'Kee & Bros., Pittsburgh, 1870s, Sprig or Ribbed Palm pattern, pressed. **(G)**
9. Pitcher, maker unknown, 1870s-80s, Strawberry Diamond and Fan pattern, cut. **(E)**
10. Pitcher, maker unknown, 1880s, white milk glass, Square Block pattern, pressed. **(F)**
11. Pitcher, maker unknown, 1880s, Mary Gregory pink enamel painting, mold blown. **(E)**

6

7

8

9

10   11

12. Pitcher, Adams & Co., Pittsburgh, 1880s, Baltimore Pear pattern, pressed. **(F)**
13. Ewer, maker unknown, 1880s-90s, satin glass with applied handle and flowers, mold blown. **(E)**
14. Pitcher, New England Glass Co., E. Cambridge, Mass., c. 1888-89, Maize pattern, pressed. **(E)**
15. Pitcher, M'Kee & Bros., Pittsburgh, 1890s, Beaded Tulip pattern, pressed. **(G)**
16. Pitcher, Co-operative Flint Glass Co., Beaver Falls, Pa., 1890s, Jewel and Dewdrop pattern, pressed. **(G)**

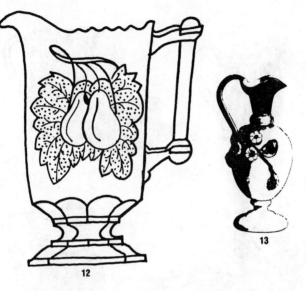

17. Pitcher, maker unknonw, 1890s, ruby and crystal, flower design, pressed. **(G)**

18. Pitcher, tumbler, and bowl, maker unknown, 1890s, crystal, pressed. **(F)**

19. Pitcher, tumblers, and tray (lemonade set), maker unknown, 1890s, engraved design, pressed. **(F)**

20. Pitcher, maker unknown, 1890s, engraved design, blown. **(G)**

21. Pitcher, maker unknown, 1890s, engraved design, pressed. **(G)**

22. Pitcher, maker unknown, 1890s, Hobnail pattern, pressed. **(G)**

23. Pitcher, T. B. Clark & Co., Honesdale, Pa., 1896, Adonis pattern, cut. **(E)**
24. Pitcher, T. B. Clark & Co., Honesdale, Pa., 1896, Jewel pattern, cut. **(E)**
25. Tankard, T. B. Clark & Co., Honesdale, Pa., 1896, Henry VIII pattern, cut. **(E)**
26. Tankard, U.S. Glass Co., Pittsburgh, 1904, Victor pattern, pressed. **(G)**
27. Pitcher, Indiana Glass Co., Dunkirk, Ind., 1928-32, Pyramid pattern, pressed. **(F)**

25. Plates

Wₜₜₕ the advent of the pressing machine in the 1820s, glass plates ranging in size from about 6″ to 10″ in diameter could be produced in large quantity. As pressing machines became more sophisticated, so did the designs invented for their molds. The first prssed plates were not usually sold as part of complete utensil sets, but were offered as separate pieces to be used for serving small cakes or other confections, to hold jam or butter, or, as in the case of cup plates (see section 9), to protect the dining table from spills and stains. The patterns available, however, usually matched other pieces available through the manufacturer, such as cups, tumblers, and bowls. Some plates in the 1800s, such as the example from Campbell, Jones & Company (no. 7), were produced as special mementoes of significant events.

Though glass plates may have been used as part of the table setting in the 1800s, it was the advent of Depression glass in the 1930s that made such pieces popular for everyday use in place of china. Companies such as Hocking Glass and Federal Glass offered complete table settings in a variety of patterns at extremely low cost. Often each pattern was available in various colors as well as in clear glass. One striking example is L. E. Smith Company's Double Shield pattern, illustrated in black (no. 20), but also available in cobalt blue, green, and pink.

1. Plate, maker unknown, 1870s, Pinwheel pattern, pressed. **(G)**
2. Plate, King, Son & Co., Pittsburgh, 1870s, Jewel pattern, pressed. **(H)**
3. Plate, Boston & Sandwich Glass Co., 1870s, Beaded Mirror pattern, pressed. **(H)**
4. Plate, Bryce, M'Kee & Co., Pittsburgh, 1870s-80s, Roman Rosette pattern, pressed. **(G)**
5. Plate, prob. Pittsburgh area, 1870s-80s, Hobnail pattern, pressed. **(H)**
6. Plate, Campbell, Jones & Co., Pittsburgh, 1870s-80s, Dewdrop with Star pattern, pressed. **(H)**

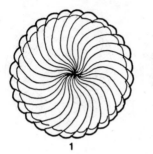

1

2

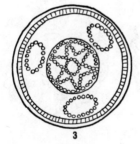

3

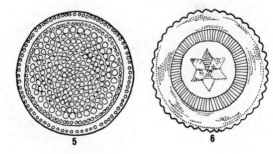

4

5

6

7. Campaign or commemorative plate, Campbell, Jones & Co., Pittsburgh, 1880-81, Garfield design, clear, pressed. **(G)**
8. Plate, Bryce, M'Kee & Co., Pittsburgh, 1880s-90s, Tulip pattern, pressed. **(F)**
9. Bread plate, prob. Pittsburgh area, 1880s-90s, Gem pattern, clear, pressed. **(H)**
10. Plate, Bryce, M'Kee & Co., Pittsburgh, c. 1888, Paneled Daisy pattern, pressed. **(G)**
11. Plate, maker unknown, late 19th century, Candlewick pattern, pressed. **(H)**
12. Plate, maker unknown, late 19th century, Jewel Band pattern, pressed. **(H)**

7

8

9

10

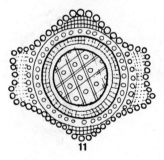

11

12

13. Plate, prob. Richards & Hartley Flint Glass Co., Tarentum, Pa., 1890s, Harvard pattern, cut. **(E)**
14. Plate, George Duncan & Sons, Pittsburgh, c. 1890, Barred Oval pattern, pressed. **(H)**
15. Plate, maker unknown, 1890s, Nellie pattern, pressed. **(H)**
16. Plate, maker unknown, 1890s, Trefoil pattern, open edge, pressed. **(H)**
17. Plate, U. S. Glass Co., Pittsburgh, 1904, Wisconsin pattern, pressed. **(H)**
18. Plate, U. S. Glass Co., Pittsburgh, 1904, Texas pattern, pressed. **(H)**

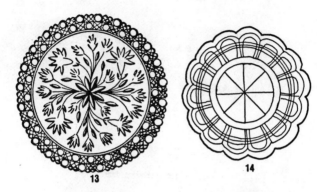

13   14

15

16

17

18

19. Plate, U. S. Glass Co., Pittsburgh, 1904, Illinois pattern, pressed. **(H)**

20. Plate, L. E. Smith Co., 1920s-1934, Double Shield pattern, black, pressed. **(H)**

21. Plate, Indiana Glass Co., Dunkirk, Ind., 1926-31, Tea Room pattern, pink, pressed. **(H)**

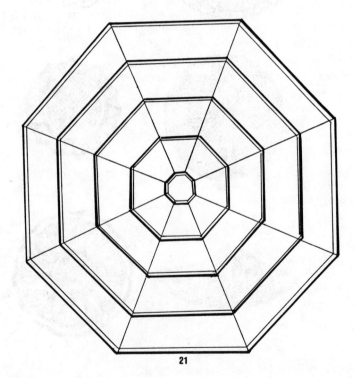

22. Plate, Jenkins Glass Co., Kokomo, Ind., late 1920s, Aunt Polly pattern, green, pressed. **(H)**
23. Plate, Hocking Glass Co., Lancaster, Ohio, 1928-30, Spiral pattern, green, pressed. **(H)**
24. Plate, Federal Glass Co., 1928-33, Rope pattern, green, pressed. **(H)**
25. Plate, Hocking Glass Co., Lancaster, Ohio, 1929-33, Block pattern, yellow, pressed. **(H)**
26. Plate, Diamond Glass Co., Indiana, Pa., 1929-31, Victory pattern, amber, pressed. **(H)**
27. Plate, Hazel Atlas Glass Co., 1930-35, New Century pattern, pink, pressed. **(H)**

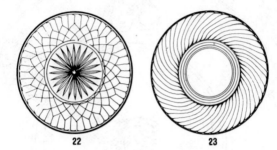

22   23

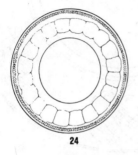 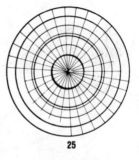

24   25

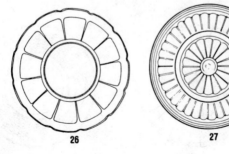

26   27

28. Plate, prob. Hazel Atlas Glass Co., 1931-33, Fruits pattern, crystal, pressed. **(H)**
29. Plate, Jeannette Glass Co., Jeannette, Pa., 1931-33, Pinwheel pattern, pink, pressed. **(H)**
30. Plate, Liberty Works, 1931-34, American Pioneer pattern, amber, pressed. **(H)**

28

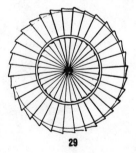

29

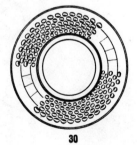

30

31. Plate, Hocking Glass Co., Lancaster, Ohio, 1934-38, Knife and Fork pattern, pink, pressed. **(H)**
32. Plate, Hocking Glass Co., Lancaster, Ohio, 1935-38, Open Lace pattern, crystal, pressed. **(H)**
33. Plate, Federal Glass Co., 1938-42, Columbia pattern, crystal, pressed. **(H)**
34. Plate, Anchor Hocking Glass Co., Lancaster, Ohio, 1939-41, Manhattan pattern, pink, pressed. **(H)**

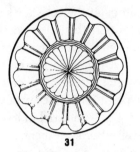

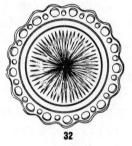

31

32

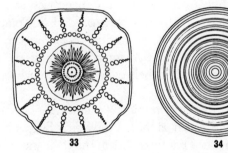

33

34

26. Platters and Trays

PRODUCTION of large glass serving dishes began to a limited extent in the mid-1800s, although the collector looking for pressed-glass trays is likely to find that these special items, when available, date largely from the last quarter of the 19th century. From the evidence of remaining dishes of the period, it seems likely that large-size serving trays and platters were infrequently made of glass. It is possible that the public was reluctant to use such a highly breakable material in large serving pieces and relied instead on silver or less expensive metals.

Some of the trays that remain were made as part of water or lemonade sets and came with two goblets or tumblers, a pitcher, and perhaps a waste bowl. Others, such as the ornate cut-glass ice cream tray made by T. B. Clark & Company (no. 10) or the imitation cut-glass celery tray offered by Fostoria Glass Company (no. 9), were intended to hold specific foods, though it is likely that their daily use was much broader in most homes.

The size of these serving pieces facilitated the creation of ambitious designs, such as that of the Lord's Supper bread tray produced by the Flint Glass Co. in the 1890s (no. 8). One imagines that the more elaborate of these pieces might never have been used on a daily basis at all, but would have been displayed in a place of honor to be admired, perhaps on a sideboard or over a mantel.

1. Cake tray, Boston & Sandwich Glass Co., c. 1835, lacy, pressed. **(F)**
2. Oval platter, Boston & Sandwich Glass Co., 1850s, Horn of Plenty pattern, pressed. **(F)**

1

2

3. Platter, prob. Richards & Hartley Flint Glass Co., Pittsburgh, 1870s-80s, Peerless pattern, pressed. **(G)**

4. Platter, Bryce, M'Kee & Co., Pittsburgh, 1880s, Sheraton or Ida pattern, pressed. **(G)**

5. Oval platter, Bryce, M'Kee & Co., Pittsburgh, 1880s, Orion Thumbprint pattern, pressed. **(G)**

3

4

5

6. Platter, Atterbury & Co., Pittsburgh, 1880s, Lion design, pressed. **(F)**

7. Tray, Duncan & Sons, Pittsburgh, 1880s, Shell and Tassel (Square) pattern, pressed. **(F)**

8. Bread tray, Model Flint Glass Co., Findlay, Ohio, 1890s, Lord's Supper design, pressed. **(F)**

6

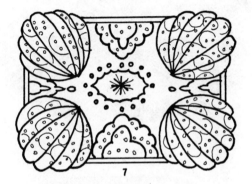

7

8

9. Celery tray, Fostoria Glass Co., Moundsville, W. Va., c. 1895, Czarina pattern, pressed. **(H)**
10. Ice cream tray, T. B. Clark & Co., Honesdale, Pa., 1896, Manhattan pattern, cut. **(F)**
11. Spoon tray, T. B. Clark & Co., Honesdale, Pa., 1896, Manhattan pattern, cut. **(F)**
12. Celery tray, T. B. Clark & Co., Honesdale, Pa., 1896, Winola pattern, cut. **(F)**

9

10

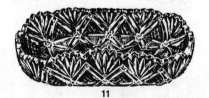

11

12

13. Celery tray, T. B. Clark & Co., Honesdale, Pa., 1896, Nordica pattern, cut. **(F)**
14. Bread plate, U. S. Glass Co., Pittsburgh, 1904, Missouri pattern, pressed. **(G)**
15. Celery tray, U. S. Glass Co., Pittsburgh, 1904, Pennsylvania pattern, pressed. **(H)**
16. Bread plate, U. S. Glass Co., Pittsburgh, 1904, Oregon pattern, pressed. **(G)**
17. Platter, Hocking Glass Co., Lancaster, Ohio, 1931-37, Mayfair pattern, pressed. **(H)**

13

14

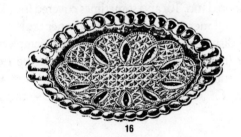

15

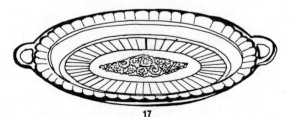

16

17

27. Salt Dishes

SALT dishes are among the most popular collectibles for admirers of early glass. Their wide variety, small size, and utility make them attractive subjects for beginning and advanced collectors alike, since they are easy to display and store and add to the attractive display on even the most modern dinner table. Glass is a perfect medium for holding salt, as other materials are easily corroded by the substance. That is the reason why silver salts are usually gilded on the inside, or provided with glass liners.

Pressed salts were among the first items widely produced by successful early-19th century American glass companies and were made available in a wide variety of shapes and patterns. Some of those patterns, such as Cable (no. 3), made to commemorate attempts at laying the Transatlantic Cable in the 1850s and '60s, were inspired by important events; others, such as Tulip and Ribbed Palm (nos. 8 and 9), imitated nature. Small salt dishes (no. 24) were intended for use at individual place settings; larger ones, termed master salts and often footed (no. 10) and sometimes covered (no. 12), could be used by all members at the family dinner table.

Salt dishes were occasionally offered complete with tiny spoons (no. 15), but in the final quarter of the 19th century, the advent of the shaker, a more convenient form, made dishes obsolete for all but the most elegant table settings.

1. Salt, Boston & Sandwich Glass Co., 1830s-40s, Lyre motif, lacy, pressed. **(F)**
2. Salt, Boston & Sandwich Glass Co., 1830s-40s, Strawberry Diamond pattern, pressed. **(F)**
3. Salt, Boston & Sandwich Glass Co., 1850s-60s, Cable pattern, pressed. **(H)**
4. Salt, M'Kee & Bros, Pittsburgh, 1850s, Harp pattern, pressed. **(G)**
5. Salt, Bryce, M'Kee & Co., Pittsburgh, 1850s, Tulip pattern, pressed. **(G)**
6. Salt, M'Kee & Bros., Pittsburgh, 1860s, Diamond pattern, pressed. **(G)**
7. Salt, M'Kee & Bros., Pittsburgh, 1860s, Imperial pattern, pressed. **(G)**

1
2

3
4

5
6

7

8. Salt, M'Kee & Bros., Pittsburgh, 1860s, Tulip pattern, pressed. **(G)**

9. Salt, M'Kee & Bros., Pittsburgh, 1860s-70s, Ribbed Palm or Sprig pattern, pressed. **(G)**

10. Salt, Bryce, M'Kee & Bros., Pittsburgh, 1870s, Diamond pattern, pressed. **(G)**

11. Salt, King, Son & Co., Pittsburgh, 1870, No. 14 pattern, pressed. **(G)**

12. Salt, King, Son & Co., Pittsburgh, 1870, No. 13 pattern, pressed. **(G)**

13, 14. Salt, maker unknown, prob. Pittsburgh area, 1880s, thumbprint motif (left), canted form (right), pressed. **(G)**, **(H)**

15. Salt, maker unknown, 1880s, Britannia metal lined with red glass, pressed. **(H)**

16. Salt shaker, maker unknown, 1880s, Hobnail pattern, pressed. **(H)**

8

9

10

11

12

13

14

15

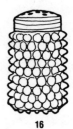

16

17. Salt shaker, Columbia Glass Co., Findlay, Ohio, c. 1888, Broken Column pattern, pressed. **(H)**
18. Salt shaker, M'Kee & Bros., Pittsburgh, 1890s, Majestic pattern, pressed. **(H)**
19. Salt dish, U. S. Glass Co., Pittsburgh, 1890s, duck form, pressed. **(G)**
20. Salt shaker, Fostoria Glass Co., Moundsville, W. Va., c. 1895, Czarina pattern, pressed. **(H)**
21. Salt shaker, maker unknown, 1890s, Parian pattern, pressed. **(H)**
22. Salt dish, maker unknown, 1890s, pressed. **(H)**
23. Salt shaker, maker unknown, 1890s, crackle motif, pressed. **(H)**
24. Salt dish, maker unknown, 1890s, Diamond pattern, canted form, pressed. **(H)**
25. Salt or pepper shaker, T. B. Clark & Co., 1896, Henry VIII pattern, pressed. **(G)**
26. Salt dish, Westmoreland Specialty Co., Grapeville, Pa., c. 1915, English Hobnail and Prism pattern, pressed. **(H)**

17

18

19

20

21

22

23

24

25

26

28. Sauce Dishes

Sauce dishes have been standard pieces in pattern glass tableware sets since the mid-1800s. The first basic form introduced has a flat bottom; later Victorian sauce dishes are often stemmed and footed. In both cases, the dish is deeper than a saucer and the sides are usually slightly curved. The diameter varies from 4" to as much as 7". Although used for serving sauce, such dishes also commonly served as ice cream and berry dishes. They might also have been used for side courses such as vegetables.

Because sauce dishes were so widely produced, collectors today can choose from an enormous variety of pressed-glass patterns and designs. The Dewdrop with Star (no. 5), Fish Scale (no. 6), Horseshoe or Good Luck (no. 7), Hobnail (no. 8), and Loop and Jewel (no. 9) patterns illustrated on the following pages are among the most commonly encountered. Other very popular patterns are Diamond Quilted, Morning Glory, Horn of Plenty, and Paneled Hobnail.

Sauce dishes come in clear glass and in many colors, including light green, amber, canary, and light blue. Frosted or shaded glass is also common. The most valuable of the dishes were made by the Boston & Sandwich Glass Co. in the 1830s and '40s and feature lacy designs.

1. Sauce dish, M'Kee & Bros., Pittsburgh, 1860s, Shell pattern, pressed. **(G)**
2. Sauce dish, King, Son & Co., Pittsburgh, 1870s, Jewel pattern, pressed. **(H)**
3. Sauce dish, King, Son & Co., Pittsburgh, 1870s, Gothic pattern, pressed. **(H)**
4. Sauce dish, King, Son & Co., Pittsburgh, 1870s, Jewel pattern, pressed. **(H)**
5. Sauce dish, Campbell, Jones & Co., Pittsburgh, 1870s, Dewdrop with Star pattern, pressed. **(H)**
6. Sauce dish, Bryce Bros., Pittsburgh, 1880s, Fish Scale pattern, pressed. **(H)**

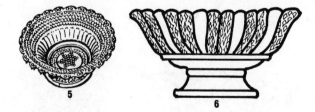

7. Sauce dish, prob. Adams & Co., Pittsburgh, 1880s, Horseshoe or Good Luck pattern, pressed. **(G)**

8. Sauce dish, maker unknown, 1880s, Hobnail pattern, pressed. **(H)**

9. Sauce dish, Alexander J. Beatty Sons, Steubenville, Ohio, 1880s, Loop and Jewel pattern, pressed. **(H)**

10. Sauce dish, Atterbury & Co., Pittsburgh, 1880s, Crossbar pattern, pressed. **(H)**

11. Sauce dish, Challinor, Taylor & Co., Pittsburgh, c. 1885, Scroll with Star pattern, pressed. **(H)**

12. Sauce dish, maker unknown, 1890s, Nellie pattern, pressed. **(H)**

7

8

9

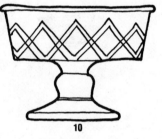

10

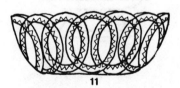

11

12

29. Spill Holders

A "spill" was a long taper of wood or strip of twisted paper, the forerunner of today's convenient fireplace matches, which was used to light oil lamps before the invention of the friction match in the early 1800s. Even after matches became popular and usually reliable, spills remained in wide use, for they were cheap, readily available, and burned longer—the last a special asset with oil lamps, which were often difficult to ignite.

The spill holder, sometimes called a spill box, case, or pot, was a footed container which held the tapers in a convenient spot near both lamp and source of flame (fireplace or stove). Enterprising glass companies produced holders to match their most popular lamp designs, and patterns such as the Waffle and Thumbprint (no. 4) can still be found on both pieces.

Today's collector will find novel uses for the spill holder—as a small flower vase or cigar holder, perhaps, or even to hold candy or after dinner mints. A pair of holders full of fireplace matches would grace any mantel, and remain closer to the original intent.

1. Spill holder, Bryce Bros., Pittsburgh, 1840s-50s, Harp or Lyre pattern, pressed. **(H)**
2. Spill holder, maker unknown, mid 1800s, Ribbed Pineapple pattern, pressed. **(G)**
3. Spill holder, maker unknown, 1850s, Smocking pattern, pressed. **(G)**
4. Spill holder, New England Glass Co., E. Cambridge, Mass., 1850s-60s, Waffle and Thumbprint pattern, pressed. **(G)**
5. Spill holder, maker unknown, Pittsburgh area, 1860s, variant of Excelsior pattern, pressed. **(G)**

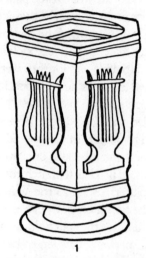
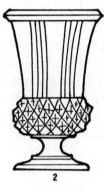
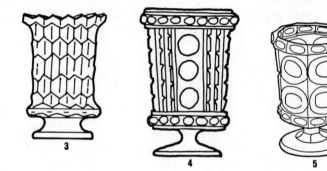

30.   Spoon Holders

A holder for spoons is a form which derived from the spill holder in the mid-19th century, although the spoon holder is usually taller and has a longer stem than the earlier form. Most often the rim is scalloped or has a sawtooth edge. A large number of spoons were needed at the typical Victorian table, and it made sense to make them available in a holder for use with various courses, or only with coffee or tea. So essential was the spoon holder that it was one of only four pieces usually included in the first tableware sets, the others being a sugar bowl, a creamer, and a butter dish.

Spoon holders can be found in most of the hundreds of pattern-glass designs popular after the Civil War, such as Harp or Lyre (no. 1), Ribbed Ivy (no. 2), and Fan and Star (no. 11). The pieces usually range in size from 4¾″ to 6″ high.

Because of the great number made, spoon holders are highly collectible objects, although they are rarely used for their original purpose in contemporary homes.

1. Spoon holder, prob. M'Kee & Bros., Pittsburgh, c. 1850, Harp or Lyre pattern, pressed. **(F)**
2. Spoon holder, maker unknown, late 1850s, Ribbed Ivy pattern, pressed. **(G)**
3. Spoon holder, Bryce Bros., Pittsburgh, 1870s, Grape Band pattern, pressed. **(H)**
4. Spoon holder, Portland Glass Co., Portland, Maine, 1870s, Loop and Dart pattern with round ornaments, pressed. **(H)**

1

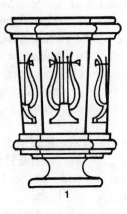

2

3

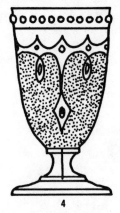

4

5. Spoon holder, King, Son & Co., Pittsburgh, 1870s, Jewel pattern, pressed. **(H)**

6. Spoon holder, King, Son & Co., Pittsburgh, 1870s, Mitchell C ware, pressed. **(H)**

7. Spoon holder, King, Son & Co., Pittsburgh, 1870s, Plain ware, pressed. **(H)**

8. Spoon holder, Adams & Co., Pittsburgh, c. 1875, Art pattern, pressed. **(H)**

9. Spoon holder, maker unknown, Pittsburgh area, late 1800s, Diamond design, pressed. **(H)**

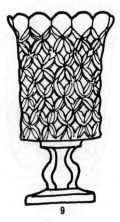

10. Spoon holder, maker unknown, late 1800s, Banded Diamond Point, pressed. **(H)**

11. Spoon holder, prob. Challinor, Taylor & Co., Pittsburgh, c. 1880, Fan and Star pattern, pressed. **(H)**

12. Spoon holder, Doyle & Co., Pittsburgh, c.1885, Triple Bar pattern, pressed. **(H)**

13. Spoon holder, prob. The Westmoreland Glass Co., Grapeville, Pa., c. 1890, Westmoreland pattern, pressed. **(H)**

14. Spoon holder, prob. George Duncan & Sons, Pittsburgh, c. 1890, Beaded Swirl pattern, pressed. **(H)**

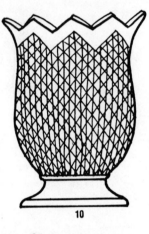

10

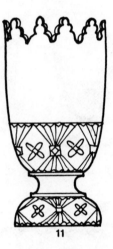

11

12

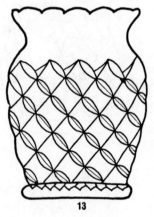

13

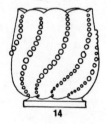

14

15. Spoon holder, Fostoria Glass Co., Fostoria, Ohio, c. 1890, St. Bernard pattern, engraved, pressed. **(H)**
16. Spoon holder, M'Kee & Bros., Pittsburgh, c. 1890, Teutonic pattern, pressed. **(H)**
17. Spoon holder, prob. George Duncan & Sons, Pittsburgh, c. 1890, Block pattern, pressed. **(H)**
18. Spoon holder, maker unknown, 1890s, crystal, pressed. **(H)**
19. Spoon holder, T. B. Clark & Co., Honesdale, Pa., 1890s, Jewel pattern, cut. **(F)**
20. Spoon holder, U. S. Glass Co., Pittsburgh, late 1800s or early 1900s, Diamond and Sunburst variant, pressed. **(H)**
21. Spoon holder, U. S. Glass Co., Pittsburgh, 1900s, Illinois pattern, pressed. **(H)**

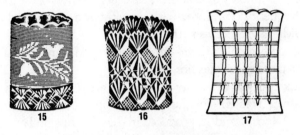

31. Sugar Bowls

AN antique sugar bowl with its original cover is one of the prizes sought by many glass collectors. Because such bowls were commonly used at the table every day, loss of the top was a frequent occurrence. With or without a cover—and some bowls made between the 1890s and 1930s were designed originally without them—an old sugar bowl is an essential part of a good glass collection.

Exceptional in beauty and design are the free-blown clear glass or colored globular or pear-shaped bowls of the early 1800s (nos. 1 and 2), produced in the New England and Mid-Atlantic states. They have a wide flaring lip and a bell-like cover. Such pleasing colors as amethyst and a cool aquamarine are frequently found.

The early pressed-glass bowls of such firms as the Boston & Sandwich Glass Co., the New England Glass Co., and various Pittsburgh area glasshouses are hardly less striking in design and execution than the free-blown forms. They often feature Gothic designs stippled in the lacy manner. The colors are brilliant—electric blue, amethyst, turquoise, and canary among them.

Each patterned glass set offered by mid- to late-Victorian makers of pressed wares included a sugar bowl and creamer. These were available in clear or colored glass. Those made of flint glass before the 1880s are generally more expensive and difficult to find than lighter bowls made according to a lime-glass formula.

1. Covered sugar bowl, Bakewell, Page & Bakewell, Pittsburgh, 1820s-30s, expanded mold blown. **(D)**
2. Covered sugar bowl, made by Charles A. Cornwall, Redwood Glass Works, S. Jersey type, 1828-48, free blown. **(E)**
3. Covered sugar bowl, prob. Bryce Bros., Pittsburgh, c. 1860s, Tulip pattern, pressed. **(H)**
4. Covered sugar bowl, maker unknown, 1860s, Lee pattern, pressed. **(H)**
5. Covered sugar bowl, M'Kee & Bros., Pittsburgh, c. 1864, Diamond pattern, pressed. **(G)**
6. Covered sugar bowl, M'Kee & Bros., Pittsburgh, c. 1864, Fairy pattern, pressed. **(G)**
7. Covered sugar bowl, M'Kee & Bros., Pittsburgh, c. 1864, Crystal pattern, pressed. **(G)**
8. Covered sugar bowl, M'Kee & Bros., Pittsburgh, c. 1864, Eugenie pattern, pressed. **(F)**

1

2

3

4

5

6

7

8

9. Covered sugar bowl, M'Kee & Bros., Pittsburgh, c. 1868, Sprig pattern, pressed. **(G)**

10. Covered sugar bowl, maker unknown, late 1800s, Helene pattern, pressed. **(H)**

11. Covered sugar bowl, King, Son & Co., Pittsburgh, 1870s, Mitchell C ware, pressed. **(G)**

12. Covered sugar bowl, prob. Doyle & Co., Pittsburgh, c. 1885, Red Block pattern, pressed. **(G)**

13. Covered sugar bowl, maker unknown, 1890s, imitation Strawberry and Fan cut glass, pressed. **(H)**

14. Sugar bowl, maker unknown, 1890s, Strawberry Diamond and Fan pattern, cut. **(F)**

15. Covered sugar bowl, maker unknown, 1890s, Nellie pattern, pressed. **(H)**

16. Sugar bowl, A.H. Heisey Glass Co., Newark, Ohio, 1890s-1900s, Hobstar or Diamond Lace pattern, pressed. **(G)**

17. Covered sugar bowl, prob. National Glass Co., Pittsburgh, c. 1900, Loop and Jewel pattern, pressed. **(H)**

18. Covered sugar bowl, U.S. Glass Co., Pittsburgh, 1900s, New Hampshire pattern, pressed. **(H)**
19. Covered sugar bowl, Hazel Atlas Glass Co., 1934-42, Wedding Band pattern, metal top, pressed. **(G)**
20. Sugar bowl, Hazel Atlas Glass Co., 1936-40, Hairpin pattern, pressed. **(H)**

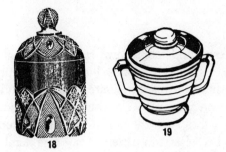

32. Tumblers and Ale Glasses

FREE-blown and mold-blown tumblers and glasses of the 1700s and early 1800s (see nos. 1 and 2) are exceptionally rare today as they were made in limited quantities and, because of their frequency of use, were easily broken. Since that time, however, it has been the practice of manufacturers to issue tumblers and glasses in pattern sets. Large numbers of these were produced, and despite breakage, the collector will find a large selection to choose from in the antiques marketplace.

Water tumblers vary in height from 3½″ to 4½″; those used for whiskey or other hard liquor are generally shorter, from 2″ to 4″. Both flat and footed forms were made throughout the 1800s; as a rule, however, footed tumblers date from early in the century. In addition to the more than 1,000 patterns issued from the 1840s until the end of the century, glass firms also turned out common home and bar ware. The usual form is simply fluted or paneled. Those of flint glass are especially valued.

Fancy sets of cut-glass tumblers for water and whiskey became very fashionable in the 1890s and early 1900s. Representative of this variety are the tumblers made by T. B. Clark & Co. (nos. 26-28) and similar wares by such firms as the Dorflinger Glass Works, Thomas G. Hawkes & Co., and the Steuben Glass Works.

Ale or beer glasses are larger vessels and in size resemble those used today, being from 6″ to 8½″ high. They are spherical in shape and have a short stem which tapers to a circular foot. Ale glasses, like the Argus and New York patterns (nos. 4 and 12 respectively), gradually fell into disuse in the late 1800s when handled mugs became popular.

1. Flip or toddy glass, Stiegel-type, Pennsylvania, late 1700s, engraved, free blown. **(D)**
2. Water tumbler, Boston & Sandwich Glass Co., 1820s-30s, sunburst with bull's eye, blown-three-mold. **(E)**
3. Tumbler, Boston & Sandwich Glass Co., 1840s-50s, Diamond Thumbprint pattern, pressed. **(E)**

1

2

3

4. Ale glass, Boston & Sandwich Glass Co., mid 1800s, Argus pattern, pressed. **(F)**
5. Footed tumbler, Richardson, Stourbridge, England, mid 1800s, crystal, cut. **(E)**
6. Tumbler, Boston & Sandwich Glass Co., early 1860s, New England Pineapple pattern, pressed. **(F)**
7. Footed tumbler, M'Kee & Bros., Pittsburgh, c. 1864, Eugenie pattern, pressed. **(G)**
8. Footed tumbler, M'Kee & Bros., Pittsburgh, c. 1868, Eureka pattern, pressed. **(G)**
9. Tumbler, M'Kee & Bros., Pittsburgh, c. 1868, Sprig pattern, pressed. **(H)**
10. Ale glass, M'Kee & Bros., Pittsburgh, c. 1868, Crystal pattern, pressed. **(G)**
11. Tumbler, M'Kee & Bros., Pittsburgh, c. 1868, Charleston pattern, pressed. **(H)**
12. Ale glass, M'Kee & Bros., Pittsburgh, c. 1868, New York pattern, pressed. **(G)**

13. Tumbler, King, Son & Co., Pittsburgh, 1870s, Argus pattern, pressed. **(G)**
14. Footed tumbler, King, Son & Co., Pittsburgh, 1870s, Maple pattern, pressed. **(H)**
15. Tumbler, Bryce Bros., Pittsburgh, late 1870s-80s, Amazon or Sawtooth Band pattern, pressed. **(H)**
16. Tumbler, maker unknown, late 1800s, Pinwheel pattern, cut. **(G)**
17. Tumbler, maker unknown, late 1800s, Strawberry Diamond pattern, cut. **(G)**
18. Tumbler, Adams & Co., Pittsburgh, late 1800s, Plume pattern, pressed. **(H)**
19. Tumbler, Bryce Bros., Pittsburgh, c. 1880s, Willow Oak pattern, pressed. **(H)**
20. Tumbler, Doyle & Co., Pittsburgh, c. 1885, Triple Triangle pattern, pressed. **(G)**
21. Tumbler, Alexander J. Beatty Sons, Steubenville, Ohio, c. 1885, No. 100 pattern, pressed. **(G)**

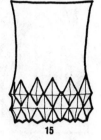

13   14   15

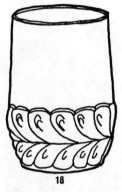

16   17

18

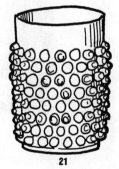

19   20

21

22. Tumbler, Riverside Glass Co., Wellsburg, W. Va., c. 1890, Riverside pattern, pressed. **(H)**

23. Tumbler, maker unknown, 1890s, Strawberry Diamond pattern, cut. **(G)**

24. Tumbler, M'Kee & Bros., Pittsburgh, c. 1894, Boston pattern, engraved, pressed. **(H)**

25. Tumbler, prob. West Virginia Glass Co., Wheeling, W. Va., c. 1894, Optic pattern, pressed. **(G)**

26. Tumbler, T. B. Clark & Co., Honesdale, Pa., c. 1896, Winola pattern, cut. **(G)**

27. Whiskey tumbler, T. B. Clark & Co., Honesdale, Pa., c. 1896, Winola pattern, cut. **(G)**

28. Tumbler, T. B. Clark & Co., Honesdale, Pa., c. 1896, Jewel pattern, cut. **(G)**

22

23

24

25

26

27

28

33. Vases

SPECIAL care has always been taken in the production of vases, whether they were intended for the most luxurious or most modest setting. As a holder for flowers, a vase had to be an object of beauty and not one of mere utility. Among the earliest glass vases produced in any quantity in America were the brilliantly colored flaring flint-glass containers pressed by the Boston & Sandwich Glass Co. in the 1840s and '50s. Most aesthetically pleasing vases, however, were free blown or mold blown rather than pressed. The glassblower could work wondrous effects.

The use of cut flowers greatly increased in the American home during the second half of the 19th century, and the market for fine vases expanded accordingly. Among the types which became available during this time were cut and blown forms (nos. 15-17) of great elaborateness, Bohemian glass imported from Europe with engraved floral designs (nos. 7-14), and various art glass forms in iridescent and opaque colors and satin and matte finishes.

At the turn of the century, vases came to be employed in home decoration primarily as art objects and only secondarily as containers for flowers. This, of course, was the era of such master craftsmen as Louis Comfort Tiffany (nos. 19 and 20), Émile Gallé (no. 18), and Frederick Carder (nos. 21 and 23).

1. Vase, maker unknown, South Jersey, late 1700s or early 1800s, tooled decoration, blown. **(C)**
2. Vase, maker unknown, English, 1850s, cut. **(D)**
3. Vase, maker unknown, 1880s, milk glass, pressed. **(H)**
4. Vase, prob. New England Glass Co., E. Cambridge, Mass., 1885, Joseph Locke, designer, cameo glass, mold blown. **(D)**

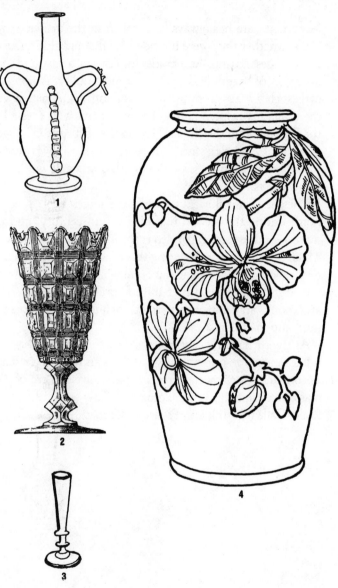

5. Bouquet holder, maker unknown, prob. Pittsburgh area, 1890s, pressed. **(H)**
6. Vase, Fostoria Glass Co., Moundsville, W. Va., c. 1895, Czarina pattern, pressed. **(H)**
7. Vase, maker unknown, European, 1890s, leaf decoration, Bohemian glass, mold blown. **(E)**
8. Vase, maker unknown, European, 1890s, floral decoration, Bohemian glass, mold blown. **(E)**
9. Vase, maker unknown, European, 1890s, floral decoration, Bohemian glass, mold blown. **(E)**
10. Vase, maker unknown, European, 1890s, floral decoration, Bohemian glass, mold blown. **(E)**
11. Vase, maker unknown, European, 1890s, floral decoration, Bohemian glass, mold blown. **(E)**
12. Vase, maker unknown, European, 1890s, floral decoration, Bohemian glass, mold blown. **(E)**
13. Vase, maker unknown, European, 1890s, floral decoration, Bohemian glass, mold blown. **(E)**

5

6

7

8

9

10

11

12

13

14. Vase, maker unknown, European, 1890s, floral decoration, Bohemian glass, mold blown. **(E)**
15. Bulb vase, T. B. Clark & Co., Honesdale, Pa., 1896, Heroic pattern, cut. **(E)**
16. Vase, T. B. Clark & Co., Honesdale, Pa., 1896, Jewel pattern, cut. **(G)**
17. Pink vase, T. B. Clark & Co., Honesdale, Pa., 1896, Winola pattern, cut. **(E)**
18. Vase, Émile Gallé, Nancy, France, c. 1898. marquetrie de verre, blown. **(B)**
19. Vase, Tiffany Decorating Co., New York, N.Y., 1896, Peacock Feather design, blown. **(B)**

14

15

16

17

18

19

20. Vase, Louis Comfort Tiffany, New York, N.Y., c. 1900, Jack-in-the-Pulpit design, Favrile glass, blown. **(C)**
21. Vase, Steuben Glass Works, Corning, N.Y., c. 1917, designed by Frederick Carder, Cintra glass, blown. **(D)**
22. Vase, Quezal Art Glass and Decorating Co., Brooklyn, N.Y., c. 1919, opalescent glass, blown. **(C)**
23. Vase, Steuben Glass Works, Corning, N.Y., 1920-30, designed by Frederick Carder, Aurene glass, blown. **(D)**
24. Vase, René Lalique, France, 1925, mold blown. **(B)**

NOTES

NOTES

NOTES

NOTES

NOTES

NOTES